Short-Circuit Withstand Capability of Power Transformers

Short-Circuit Withstand Capability of Power Transformers

Zur Erlangung des akademischen Grades eines

DOKTOR-INGENIEURS

an der Fakultät für
Elektrotechnik und Informationstechnik
des Karlsruher Instituts für Technologie (KIT)
genehmigte

DISSERTATION

von

Dipl.-Ing. *Daniel Hermann Geißler*
geb. in: Bruchsal, Deutschland

Tag der mündlichen Prüfung: 21. Juli 2016
Hauptreferent: Prof. Dr.-Ing. Thomas Leibfried
Korreferent: Prof. Dr.-Ing. habil. Thomas Seelig

Bibliografische Information der Deutschen Nationalbibliothek
Die Deutsche Nationalbibliothek verzeichnet diese Publikation in der Deutschen Nationalbibliografie; detaillierte bibliografische Daten sind im Internet über http://dnb.d-nb.de abrufbar.

1. Aufl. - Göttingen: Cuvillier, 2016
Zugl.: Karlsruhe (KIT), Univ., Diss., 2016

Nonnenstieg 8, 37075 Göttingen
Telefon: 0551-54724-0
Telefax: 0551-54724-21
www.cuvillier.de

1. Auflage, 2016
Gedruckt auf umweltfreundlichem, säurefreiem Papier aus nachhaltiger Forstwirtschaft.

ISBN 978-3-7369-9332-7
eISBN 978-3-7369-8332-8

Acknowledgments

This doctoral thesis comprises results from my research project dealing with the short-circuit performance of power transformers. It has been written during my work as a research associate at the Institute of Electric Energy Systems and High Voltage Technology (IEH) of the Karlsruhe Institute of Technology (KIT).

I would first of all like to thank my doctoral supervisor Professor Dr.-Ing. Thomas Leibfried for his encouraging guidance, assistant comments and suggestions during the course of this work. The trusting relationship and the opportunity to contribute and pursue my own ideas with his full support deserves particular mention.

I am very grateful towards the co-examiner Professor Dr.-Ing. habil. Thomas Seelig for his friendly interest in this thesis. The very valuable discussions, his suggestions and comments from the viewpoint of mechanics decisively improved this work.

Special gratitude goes towards my brother Christian for his great support in the field of CAD modeling and mechanical simulation. He and his wife Julie deserve many thanks for their assistance with my written English in the present thesis and publications over the last years.

I am also indebted to all my undergraduate students for the trustworthy collaboration and their energetic support, especially regarding the numerous measurements. This doctoral work would not have reached the present extend without their committed work during their theses.

The kind and countless assistance of the institute's electrical and mechanical workshop is greatly appreciated. Without their support, the short-circuit forces test stand could not have been brought to fruition in the present form. The staff of the secretariat deserves heartfelt gratitude for their kindness, cheerfulness and the reliable support in administrative tasks.

A particular debt of gratitude is owed to the Siemens AG transformers plants Nuremberg and Weiz for their friendly cooperation and for manufacturing the iron core of the test stand. The ASTA Elektrodraht GmbH deserves many thanks for providing many meters of test conductors and the valuable suggestions and comments on the field of Continuously Transposed Conductors.

Last but not least my gratitude extends towards all my past and present colleagues of the Institute of Electric Energy Systems and High Voltage Technology (IEH) for the strong team spirit and the pleasant and trustworthy working environment. Special thanks go to Christian Freitag and Michael Suriyah for their valuable suggestions and the profitable discussions in the field of power transformers.

Karlsruhe, in July of 2016

Daniel H. Geißler

Contents

Zusammenfassung (German Abstract)

Im Rahmen der vorliegenden Arbeit wurde die Kurzschlussfestigkeit von Leistungstransformatoren untersucht. Der Schwerpunkt liegt auf dem Fehlerbild des radialen Einknickens von Teilen der Wicklungen dem sogenannten Buckling. Die Ursachen sind hohe Stromkräfte, welche die Festigkeit der Wicklungen übersteigen, oder eine alterungs-bedingte Abnahme der Widerstandsfähigkeit der Wicklungen gegen solche Effekte. Hohe Fehlerströme entstehen oftmals durch Transformatornahe Kurzschlüssen im Energienetz oder durch erhöhte Einschalt-Ströme (Inrush-Effekt).

Ein Fokus dieser Arbeit liegt auf der mechanischen Charakterisierung verschiedener Leitertypen für Transformatoren. Besonders der Drilleiter mit seinem komplizierten Aufbau aus vielen Einzelleitern ist in der Literatur bisher noch nicht ausreichend untersucht worden. Dieser Leitertyp wird jedoch in einer Vielzahl von großen Netzkuppel- und Maschinentransformatoren eingesetzt, um die Verluste durch Wirbelströme und Kreisströme in den Wicklungen zu reduzieren. Der grundlegende Ansatz für die analytische und simulations-basierte Beschreibung von Drilleitern basiert auf einer Vernachlässigung der Transpositionen zwischen den Einzelleitern. In dem daraus resultierenden, vereinfachten Modell befinden sich die Einzelleiter stets auf derselben, radialen Position in Bezug auf die Wicklung, was einem einfachen, gestapelten Aufbau entspricht. Dieses Modell konnte mittels Finite Elemente basierten Strukturanalysen, sowie experimentellen Test als ausreichend genau bestätigt werden.
In vielen modernen Transformatoren werden häufig verklebte Drilleiter eingesetzt. Durch eine thermisch aktivierbare Klebeschicht zwischen den Einzelleitern erhöht sich die Steifigkeit dieses Leitertyps signifikant. Diese Arbeit stellt ein Simulationsmodell für solche Leiter, basierend auf einem Kohäsiven-Elemente-Ansatz vor. Die dafür notwendigen Parameter stammen aus Scherzug- und Schälzugversuchen an verklebten Einzelleitern. Das durch diese Tests parametrisierte Kohäsive-Elemente-Modell konnte anhand von Biegetests verifiziert werden. Es erlaubt eine präzise Analyse der Belastung in den Klebeverbindungen zwischen den Einzelleitern unter dem Einfluss von Stromkräften auf eine Wicklung.
Weitere Untersuchungen wurden zum Einfluss der Isolierpapier-Alterung im Transformator auf die Knickfestigkeit von nicht verklebten Drilleiter-Wicklungen durchge-

führt. Mittels eines Versuchsstandes, der die Kräfteverhältnisse aus einem Leistungstransformator realistisch auf eine einfache Testwicklung abbilden kann, wurde die dynamische Knickfestigkeit von papier-isolierten Drilleitern mit und ohne Isolation untersucht. Der letztere Fall soll einer drastisch gealterten Isolation entsprechen, welche weitestgehend nicht mehr vorhanden ist. Für die untersuchten Leitertypen konnten Zunahmen in der Knickfestigkeit als Resultat der Papier-Isolation von bis zu 40 % ermittelt werden. Statische Biegetests ergaben leicht geringere Zunahmen in der Knickfestigkeit verglichen mit den dynamischen Tests waren jedoch insgesamt vergleichbar in der Aussage. Biegeversuche an weiteren Leitertypen, die nicht im dynamischen Versuchsstand untersucht werden konnten, ermittelten Zunahmen um bis zu 200 %. Diese Untersuchungen verdeutlichen den signifikanten Beitrag der Isolation zur Steifigkeit der nicht verklebten Drilleiter. Eine genauere Auswirkung der Alterung konnte mittels beschleunigt gealterten, papier-isolierten Leiter-Proben mittels Biegeversuchen gezeigt werden. Hier zeigte sich, dass selbst eine stark gealterte Isolation noch eine gewisse Festigkeit besitzt. Unabhängig von der Beeinflussung der mechanischen Festigkeit des Leiters ist auch von Interesse ab wann die Isolation bereits geringfügig beschädigt wird, da damit deren elektrische Festigkeit abnimmt. Ein Simulationsmodell basierend auf den mechanischen Eigenschaften der gealterten Papierproben ermöglicht eine genauere Analyse zum Einsetzen von Isolationsschäden.

Ein weiterer, wichtiger Teil dieser Arbeit widmet sich analytischen Beschreibungen und simulations-basierten Untersuchungen zum Buckling von Wicklungen. Eine analytische Formel zur Berechnung der kritischen Knicklast von Drilleitern unter Berücksichtigung von elastisch-plastischem Materialverhalten konnte mittels Simulationsmodellen verifiziert werden. Weitergehende Simulationen einer kompletten Windung mit radialen Abstützungen zeigten, dass die Steifigkeit der Lagerung einen entscheidenden Einfluss auf die Knickfestigkeit der Wicklung hat. Ein Korrekturfaktor für die analytische Beschreibung, der nur von der Steifigkeit der Abstützungen abhängt, wurde aus dem Simulationsmodell ermittelt. Ohne diesen Faktor sind die Ergebnisse jedoch als sichere Abschätzung anzusehen, da die Knickfestigkeit zu niedrig bestimmt wird. Die dynamischen Knick-Versuche mithilfe des Versuchsstands konnten sowohl analytisch als auch simulations-basiert verifiziert werden. Da die alle Tests, Simulationen und Gleichungen den Einfluss des im Transformator vorhandenen Isolieröls vernachlässigen, wurde eine Beschreibung für dynamische Knickvorgänge mit dem Effekt der Öl-Dämpfung untersucht. Ein signifikanter Einfluss des Isolieröls auf die kritische Knicklast konnte nicht ermittelt werden. Einzig der reduzierte Reibungskoeffizient zwischen den Einzelleitern ist eine geringfügige Einflussgröße.

Die vorgestellten Ergebnisse und Simulationsmodelle finden im Bereich der Transformator-Auslegung Anwendung. Die vorgestellten, analytischen Gleichungen und Vereinfachungen konnten mittels Simulationen als sichere Abschätzung nach unten hin verifiziert werden.

Chapter 1

Introduction

Power transformers are a critical and essential part of each power transmission system. They are used as generator step-up transformers inside power generation plants, as substation transformers to interconnect power grids with different voltage levels, or as distribution transformers to connect households to the power grid, just to name a few essential applications. Even modern developments like HVDC links or the connection of offshore wind farms to the onshore grid depend on the voltage transformation capability of power transformers.

Nowadays, utilities and network operators are interested in maximizing cost-effectiveness of each piece of equipment installed in the grid. This focus is a result of advancing deregulation of energy markets and the related increase in competition. Especially power transformers, which rank among the most expensive components of power transmission and distribution systems are in focus. In conclusion, the early detection of transformer faults plays an important role in cutting potential transformer damages and in lowering the risk of unexpected power outages. Aggravating this situation, during the last decade, a large percentage of the world's transformers population has reached or exceeded its expected lifetime. Utilities and network operators working under growing cost-effecting pressure, therefore, show an increased interest in transformer life evaluation and monitoring. The aim is to decide whether an aged transformer is still fit to operate, or needs to be retrofitted or even replaced. For this purpose, modern diagnostic measurements like frequency response analysis, dielectric spectroscopy, winding resistance test or gas-in-oil analysis are used. In [66] the authors provide more details on condition monitoring and diagnosis methods.

Apart from aging processes, various faults arising during normal operation can impact transformer reliability and lifetime. Some examples are high mechanical stress caused by magnetic forces due to high short-circuit currents, dielectric stress resulting from high impulse voltages or partial discharge as effect of deteriorated transformer insulation.

The steadily increasing demand for electrical power results in growing generating capacities and increasing interconnections in power grids. These developments contribute to an increase in short-circuit capacity of power grids with negative consequences, especially for power transformers. Rising short-circuit currents in such strongly connected grids make the short-circuit withstand capability of power transformers more severe.
Under normal operation conditions transformer windings are exposed to low magnetic forces. In a short-circuit event winding currents can increase up to 10...30 times based on rated current [17]. As magnetic forces depend on the square of the current, they rise by a factor of 100...900. Although protection devices usually quickly interrupt fault currents after a few supply voltage cycles this time is long enough to cause serious damage to the transformer. A further downside is that the peak short-circuit current occurs within the first cycle [36, p. 239] and can take values up to twice of the steady-state short-circuit current. The occurrence of short-circuit events in power grids is quite rare but the probability of such events increases during the transformer lifetime which is typically up to 50 years.

The result of a transformer failure survey based on the investigation of 750 massive transformer failures from 1996 to 2010 confirms the importance of short-circuit strength [59]: Almost half of the failures have been correlated to windings and the insulating system, as shown in Figure 1.1 a).
A low short-circuit strength can result in mechanical deformation of windings, breakdown or failure of clamping structures which may end up with winding-winding or winding-ground fault connections. Figure 1.1 b) shows the effect of excessive short-circuit forces on a transformer winding which failed by radial buckling. Such internal faults can end up in bursting of the tank, blow-out of bushings or even fire. In short, insufficient short-circuit strength can lead to a total loss of the transformer, and additional severe environmental impact.
Windings and supporting structure are designed to withstand short-circuit forces without permanent deformation or failure. However, aging processes such as cellulose degradation can favor the initiation of these faults. Insulating system materials, which often serve as winding support structure experience a reduction in stiffness and mechanical strength due to aging.
However, also factory new power transformers are not immune to winding faults. Especially, the short-circuit behavior of large power transformers, which today cannot be tested, is highly topical.

a)

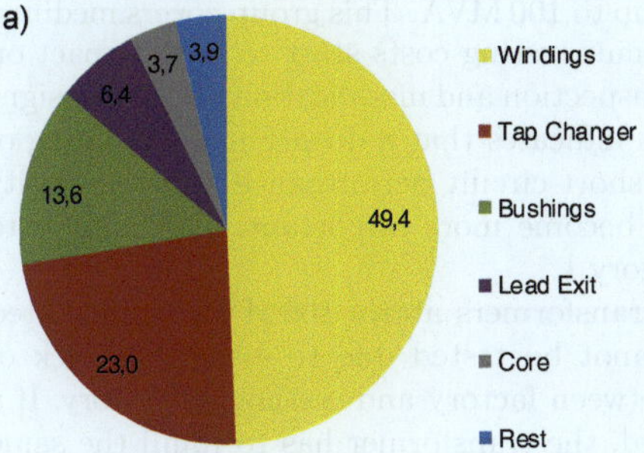

b)

Figure 1.1: a) Fault location distribution from transformer failure survey [59]
b) Deformed power transformer winding resulting from high short-circuit currents

1.1 Standardization of Short-Circuit Withstand Capability

The short-circuit strength is among the most challenging aspects of transformer design. This section gives a brief insight into the underlying standard IEC 60076-5 [68] with focus on 3-phase core-type transformers with separate windings. A power transformer has to resist both thermal and mechanical effects of short-circuit events. Thermal withstand capability has to be demonstrated by calculation as testing would lead to degradation of the transformer. The standard provides equations for calculating the average temperature of each winding during a symmetric short-circuit with a duration of 2 s as a worst-case. For oil-immersed transformers with insulation system of thermal class A (105 °C) the greatest permissible temperature of a copper winding is 250 °C.
The ability to withstand mechanical effects of short-circuits can be demonstrated either by tests or by calculation and design considerations. For this purpose the standard defines three transformer categories:

- Category I is up to 2.5 MVA and covers small distribution transformers. The short-circuit performance validation is usually done by testing of one or two units. This is the cheapest and fastest way to verify conformity. If a transformer does not pass testing, it is easy to do the disassembly and failure analysis. Improvements and modifications are often cheap compared to order costs. The short-circuit test is passed if a visual inspection of the active part does not show changes in comparison to its original condition before the test. Measurement of short-circuit impedance after the test is another important criterion. Its value may not differ by more than 2 % from the original value. Furthermore, routine tests including dielectric tests have to be successfully repeated after short-circuit testing.

- Category II is above 2.5 MVA and up to 100 MVA. This group covers medium-size power transformers. Short-circuit testing costs start to have impact on the total order expenditure. Also, inspection and modification of failed designs require much more resources. This indicates that a deeper understanding of design parameters that influence short-circuit performance and the ability to forecast short-circuit behavior become more important. The criteria to pass short-cirçuit test are of category I.
- Category III covers large power transformers above 100 MVA. Units rated above several 100 MVA often cannot be tested due to either the lack of test facilities or large distances between factory and testing laboratory. If a short-circuit test can be conducted, the transformer has to fulfill the same conditions as for category I, except the limit of change in impedance which is tightened to 1 %. Otherwise, if testing is not practical, a critical design review of manufacturer's calculations has to be carried out.

Particularly concerning large power transformers with ratings above 100 MVA it is essential that manufacturers can calculate and predict their short-circuit performance precisely. This demand is associated with the need for precise design and simulation tools, a reliable material characterization and experienced engineers to calculate the short-circuit behavior of power transformers.

1.2 Thesis Objectives

1.2.1 Current State of Science

In the past, there has been a variety of experimental, analytic and simulation-based work about the short-circuit strength of power transformers: In [54] the authors presented a short-circuit test apparatus to measure the strength of inner transformer windings against radial forces. [38] proposed a study about short-circuit strength of large transformer windings. The authors derived a simple equation for the radial buckling strength of windings based on a circular ring. The authors of [34] published further investigations including the effect of supporting spacers and material plasticity. Another experimental study of [57] conducted radial buckling tests on small transformer winding models. With these tests, the effects of various types of radial support structures on the radial buckling phenomena have been investigated and analyzed by the means of analytic calculations including nonlinear material behavior. In [65] dynamic simulations regarding the axial and radial deformation behavior of multilayer disk-type transformer coils are presented. The authors of [7] calculate spiraling effects of conductors in a helical transformer winding based on a 3D simulation model. In [5] the effects of winding process-induced strain on the mechanical strength of transformer windings exposed to radial short-circuit forces are investigated.
This little insight into the state of science points out that the short-circuit performance of power transformers is a long-standing traditional topic. The availability of new simulation tools and a steadily increasing computation power open up

new and more precise calculation approaches. A variety of simulation work has already been done in the field of short-circuit current and forces analysis [26, 37, 21, 1]. However, there hardly exist publications about the mechanical effects of these forces and specially how to handle stranded conductors like the Continuously Transposed Conductor (CTC). Especially this kind of conductor has become increasingly popular for the use in large power transformers over the last couple of years, as it significantly reduces losses.
In [49] tilting failures of CTCs due to axial short-circuit forces have been investigated. Concerning the strength against radial forces, the author of [17, p. 275] proposes to use a model with an equivalent thickness of twice the strand thickness without further proofs. In the case of bonded CTCs the same book assumes a thickness equal to 80 % of the outer CTC dimension in the radial direction.
However, the following quotation intensifies the need for more investigation, in particular on the topic of CTCs:

> "An area of some uncertainty is how to treat transposed cable, with or without bonding, in the stress calculations. It is not exactly a solid homogeneous material, yet it is not simply a loose collection of individual strands. We believe that we have taken a conservative approach in our calculations. However, further experimental work in this area will be useful." [17, p. 313]

1.2.2 Buckling Analysis on Transformer Windings

This work presents analytic equations for radial buckling phenomena adapted to CTC-based, inner transformer windings. Furthermore, simulation models are proposed based on Finite Element Analysis (FEA). A comparison between analytic and simulation-based approaches points out the limitations and influencing factors of the analytic equations.
Experimental work has been conducted in the field of dynamic short-circuit testing by the means of a test stand. The aim was to expose test windings to similar magnetic force conditions as in the short-circuit case of a real transformer. The development of acceleration sensor and camera based deformation measurement systems for the use within this test stand are explained. The winding's buckling strength gathered from the experiments is compared to the analytic and simulation-based approaches with good correlation.

1.2.3 Characterization of Conductors

Precise calculations and simulations demand for reliable material characterization. Therefore, the necessary parameters for copper as a main material have been estimated from tensile tests. The insulating paper properties were characterized by the means of zero- and wide-span tensile tests. A precise FEA simulation of bonded CTCs is possible based on adhesive parameters deduced from shear- and peel-tests. Bending tests verify the results from the analytic and simulation based CTC models.

1.2.4 Impact of Insulating Paper Aging

Static bending tests on bare and paper insulated conductors enable preliminary investigations of the impact of the paper covering on the overall conductor stiffness. Dynamic tests on the developed test stand confirm this stiffness contribution of the paper insulation. The contribution of the insulating paper to the total conductor stiffness is expected to be largest for unbonded CTCs: Due to thin strands, the stiffness of the copper core is relatively small while the stiffness of the paper covering significantly increases with the outer CTC dimensions.
The investigation of aged conductors from old transformers is not practical as it is very difficult to obtain aged test samples from real transformers. To get around this, aged conductor samples have been produced by the means of an artificial aging procedure. Due to the length limitation for the aged samples, a testing on the test stand has not been possible. However, the bending tests conducted with aged, paper covered CTC samples demonstrate how the cellulose deterioration affects the buckling strength of such conductors. With the help of simplified simulation models, a more detailed analysis is possible based on the measured stress-strain characteristic for the aged insulating paper.

Chapter 2

Fundamentals

2.1 Power Transformer Windings

2.1.1 Winding Types

In the most typical case of a three phase core-type power transformer, each limb of the core contains at least two windings named Low Voltage (LV) winding and High Voltage (HV) winding. In the case of a variable transformation ratio the tappings usually are connected to the HV winding. The corresponding turns are separated in the so-called tapping winding. Some high rating transformers often have a third tertiary winding, apart from the LV and HV winding. The tertiary winding is connected in delta formation, and limits fault currents in the event of a short-circuit between line to neutral. During normal operation, it reduces the unbalancing in the LV winding in the case of an unbalanced load. Sometimes the tertiary winding supplies auxiliary loads at a different voltage level than the HV load. Figure 2.1 visualizes a typical winding arrangement by the means of cut-through views of a 5 limb 3 phase core.
Power transformer windings are wound as concentric cylinders. As the core provides the flux return path, it encloses the windings. One principal aim in transformer design is to reach a low leakage flux in the core windows to reduce winding eddy current losses. Generally, tall and slim windings have less leakage flux compared to a short, squat design [24, p. 123]. In a tall and slim design, the leakage flux will be mostly axially directed. Therefore, it is essential to divide the conductors in the direction perpendicular to the leakage flux (the radial direction) to reduce eddy current losses. The subdividing strands have to be individually insulated against each other.

The two most important winding types for power transformers are layered and disk winding, as shown in Figure 2.2.

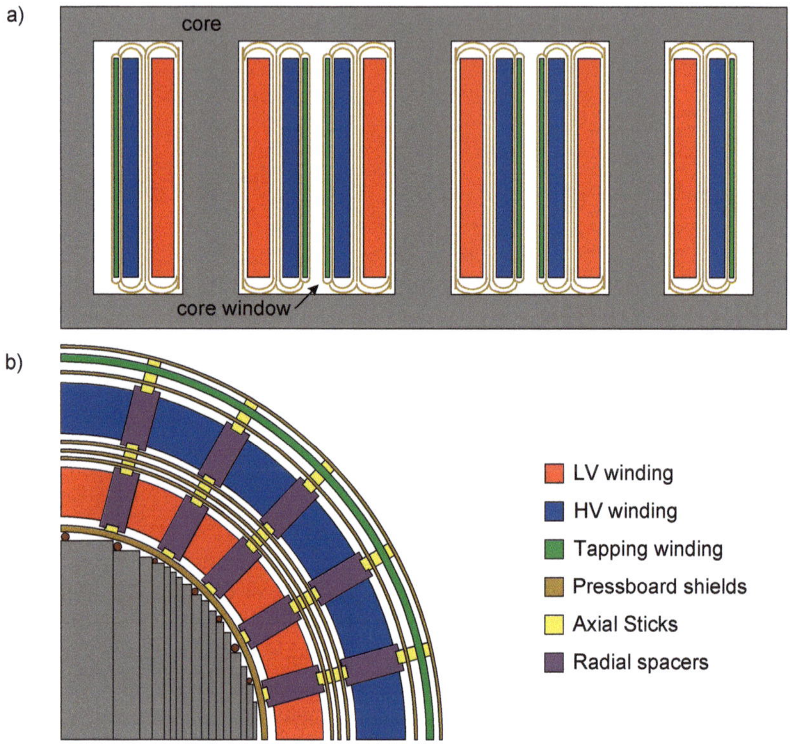

Figure 2.1: a) Cut-through a 5 limb 3 phase core
b) Cut-through view of one phase

The layered type is usually used for the LV winding. Due to its low voltage rating, the LV winding is located close to the grounded core. It is wound in vertical layers (top–bottom, bottom–top, ...) on a robust tube of high strength insulating material like synthetic resin bonded paper. Between the tube and the innermost winding layer axial sticks made from Pressboard are placed, to provide axial cooling ducts for the transformer oil. If necessary, radial spacers between the turns in axial direction provide radial cooling ducts.
LV and HV winding are insulated from each other by several Pressboard cylinders separated by oil gaps. This method provides a high dielectric strength barrier. Axial sticks support the cylinders against each other, providing a rigid support for the HV winding, which is usually built in disk type construction.
Disk type windings are arranged in horizontal layers called disks. The winding sequence is alternately out–in, in–out, and so on. The turns within a disk either touch each other, or are separated by Pressboard spacers. This method provides axial ducts to improve cooling performance. Furthermore, structural separators in

the form of radial spacers form radial cooling ducts and support the whole winding in the axial direction. As the voltage drop between two adjacent disks is quite moderate, compared to the layers of a layered winding, the disk type winding is preferred for HV windings. Another advantage concerning high voltage surges is the possibility to interleave disks near the HV lead exit. This method influences stray capacitances between disks. Based on this method, it is possible to shape the surge voltage distribution along the winding height.

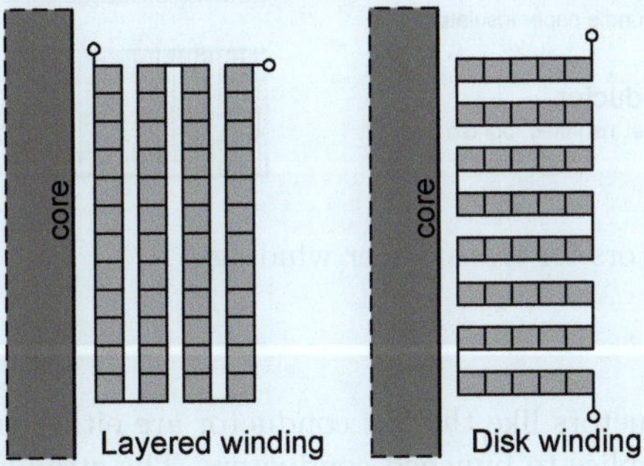

Figure 2.2: Two major types of windings for core-form power transformers

2.1.2 Conductor Types

Medium and large power transformer windings are usually made from high-conductivity copper. The most commonly used conductor is the rectangular shaped conductor, as it offers the best space factor for a winding. The latter defines the ratio between copper cross-section and total winding cross-section. Circular cross-section wires decrease the space factor, as there are many gaps between the turns. Furthermore, the mechanical stability of a winding made from round wires is lower, compared to rectangular wires. Therefore, circular cross-section wires are restricted to distribution transformer sizes [24, p. 60]. Wires with a rectangular cross-section can be single stranded flat conductors, bunched conductors or Continuously Transposed Conductors (CTC). Some schematic examples are drawn in Figure 2.3. Special conductor types, usually used in low voltage coils of distribution transformers, are copper or aluminum foils.

Insulating materials directly applied to the conductor provide electric insulation between winding turns. Wide-spread methods are enamel-coating of conductors or a paper insulation. The enamel itself is often used as polyvinyl acetate coating with a thickness ranging from 50 ... 70 µm. Its principal purpose is to insulate multi-stranded conductors against each other. If a higher dielectric strength for turn-turn insulation is needed, the paper covering is the preferred method. More details concerning paper insulation will be given at the end of this section.

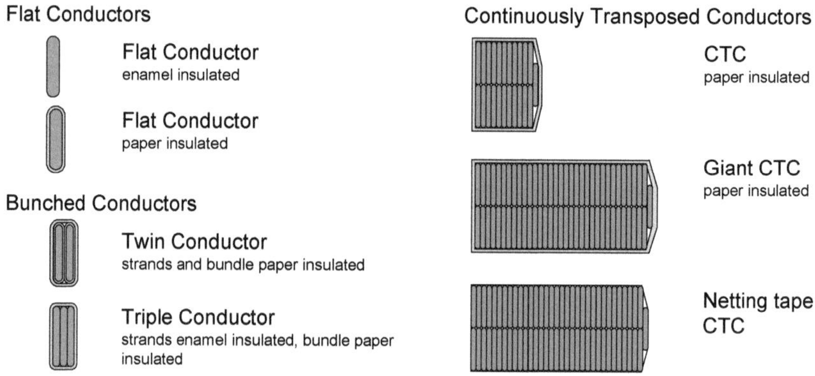

Figure 2.3: Conductors for transformer windings

Single-stranded conductors like the flat conductor are either enamel or paper insulated. The same applies to bunched conductors. The strands are often enamel insulated with an overall paper covering around the whole bundle. There are also conductors with individual paper covering around each strand and a bundle paper covering.

CTCs consist of many enamel insulated strands, which are arranged in two stacks. Usually, there is also a paper covering around the whole bundle (Figure 2.4 a)). The advantage, which arises from the use of CTCs, is a significant reduction in eddy current losses due to the small strand size. The transposition of the strands (Figure 2.4 b)) compensates loop voltages. Such residual voltages are induced between the strands connected in parallel, due to a variation of stray magnetic field with the radial winding position. The distance between two transpositions has to ensure, that each strand has passed through the same average stray magnetic field within one winding turn. Another advantage is the reduction of winding time and the reduced space factor for CTC-based windings. The higher amount of copper inside the core window compared to insulating materials facilitates a reduction of core dimensions and overall transformer dimensions, thus reducing its costs.
For the purpose of a high short-circuit forces withstand capability, thermosetting epoxy bonded CTCs are available. Their strands are covered by a thermosetting epoxy resin. This resin polymerizes during the drying process of the active part, resulting in a significantly increased bending stiffness of the whole CTC.
Depending on the dielectric strength needed between the turns of a coil, different types of electrical insulation are used. The most commonly used material is Kraft paper. It is among the cheapest and best electrical insulation materials known [24, p. 61]. The paper is produced in strips with a width ranging from 20 . . . 30 mm.

Insulating paper is an anisotropic material with different mechanical properties in the manufacturing direction (MD), compared to the cross direction (CD). The latter corresponds to the strip width. During production the cellulose fibers orient in machine direction, resulting in a higher tensile strength and stiffness [62]. Usually, the paper strips spirally coat the conductors. The number of layers depends on the desired dielectric strength. Up to three different wrapping methods can be combined for the insulation of a CTC (Figure 2.4 c)). The aim is to reduce bulging of the outermost layers, as this will constrain oil circulation and therefore limit the cooling performance. For special applications there are further paper types with particular properties [24, p. 65]:

- Creped paper is a crimped paper with significantly increased extensibility in manufacturing direction. This property allows wrapping of irregular elements, like lead exits or tapping leads, without bulging or sticking out. One disadvantage of creped paper is the loss of elasticity over time.
- Highly extensible paper has enhanced mechanical properties like a high burst and tear strength. This kind of paper is best suited for the automated covering of electrostatic stress rings or large cross-section CTCs. Especially the latter is difficult to bend to the desired winding radius, inducing severe stress in the insulating layers or creating bulging and wrinkling.
- Thermally upgraded paper provides better temperature stability, thus reducing thermal degradation. To achieve these properties, stabilizers are added during the paper manufacturing process.
- Diamond dotted paper is coated with thermosetting resin dots on one side. When the transformer's active part is heated for drying purposes, the resin melts and cures. This process strongly enhances the mechanical strength of the insulating structure. However, the drying and impregnation properties of the paper are negatively affected by the resin dots.
- Especially for CTCs with higher cooling efficiency a covering with insulating paper is not recommended. The thermal resistance between the copper strands and the oil is much lower without paper. Nevertheless, to hold the strands together, a so-called netting tape is used. It is made from polyester or mixed polyester glass fiber yarns which are woven to a wide-mesh tape.

Nowadays it is a trend to use paperless CTCs especially for low voltage (LV) coils of large power transformers, as they increase the cooling capability compared to paper covered conductors. As a result, the LV winding can operate at a higher current density. Bulging of paper between radial spacers, which could restrict the oil flow, does not happen. It follows that radial cooling ducts of paperless CTC coils can be around 1 ... 1.5 mm smaller than in coils made from paper insulated CTCs [51]. Disadvantages of paperless CTCs are their lower impulse voltage withstand strength for same clearances compared to paper insulated CTCs. The electric stress at the interface between paper and oil is less than the stress on the surface of a bare or enamel covered conductor in oil, due to the higher permittivity of oil impregnated paper (3.3) compared to transformer oil (2.2) [51]. Therefore, the electric field is displaced from materials with higher permittivity. A paper insulation increases the edge radii which further reduces electrical stress. Due to these reasons, paperless CTCs are mainly used for windings up to 36 kV. Care has to be taken during the winding process, to not damage the enamel coating. The detection of damaged insulation may be more difficult in this case than it is for paper covered CTCs.

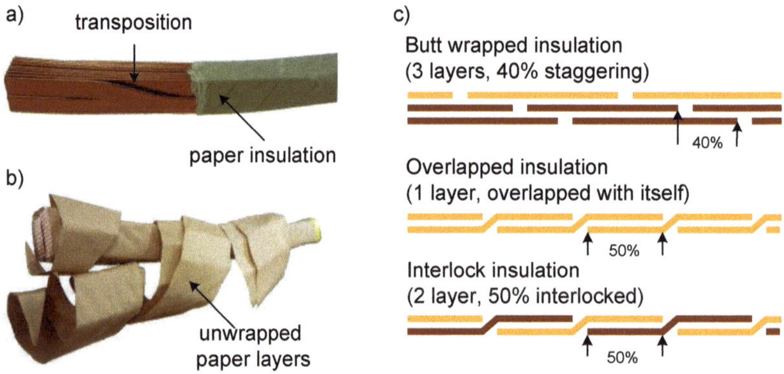

Figure 2.4: a) Paper insulated CTC, b) CTC with partially unwrapped paper insulation, c) Insulating paper wrapping methods

2.1.3 Copper for Electric Applications

The copper used for electrical conductors is usually high conductivity electrolytic-tough-pitch (ETP) copper or oxygen-free high conductivity (OFHC) copper. The latter is the preferred metal for the manufacture of power transformer conductors, due to its much lower oxygen content and the resulting increased conductivity compared to ETP copper. OFHC copper contains a maximum of 10 ppm of oxygen compared to 200 … 600 ppm for ETP.
Different grades in yield strength can be produced by cold work, which usually is a rolling process for electrical wires and conductors. According to [3] a rolling reduction by 70 % increases the tensile strength of ETP copper by more than 70 %. The Young's modulus also varies between 110 GPa for annealed and 130 GPa for cold worked ETP copper.
However, cold working reduces the conductivity as it elongates the grains in the copper. According to [3] a reduction in conductivity ranging from 1 … 5 % depending on the rolling reduction is possible. A final annealing process restores the conductivity, as it forces new grains to recrystallize, but the high strength properties are lost.
Therefore, a careful measurement of the mechanical properties characterizing the elastoplastic behavior of copper is inevitable for precise calculations.

2.2 Short-Circuit Considerations

2.2.1 Short-Circuit Current

The single-phase equivalent diagram of a transformer in case of a short-circuit consists of transformer short-circuit impedance $R_t + iX_t$ and the feeding network impedance $R_g + iX_g$, as shown in Figure 2.5. This leads to a resistive-inductive circuit with total impedance $R_k + iX_k$, feeding voltage $u_0(t) = U_0 \sin(\omega_0 t + \vartheta)$ and short-circuit current $i_k(t)$.

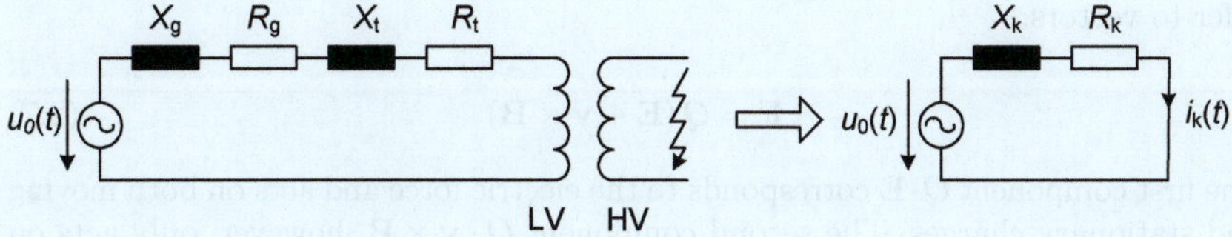

Figure 2.5: Power transformer single-phase short-circuit equivalent diagram

If a short-circuit appears at the time $t = 0$, the voltage $u_0(t)$ has the phase angle ϑ. According to [12, p. 15] the short-circuit current $i_k(t)$ is given by the following equations with peak short-circuit current $\hat{I}_k$, phase angle φ_k of the circuit impedance and circuit time constant τ_k:

$$i_k(t) = \hat{I}_k \left[\sin(\omega_0 t + \vartheta - \varphi_k) - e^{-t/\tau_k} \sin(\vartheta - \varphi_k)\right] \tag{2.1}$$

$$\hat{I}_k = U_0/\sqrt{R_k^2 + X_k^2} \tag{2.2}$$

$$\tan\varphi_k = X_k/R_k \tag{2.3}$$

$$\tau_k = X_k/(\omega_0 R_k) \tag{2.4}$$

The short-circuit current $i_k(t)$ has a sinusoidal and an exponentially decreasing component. The steady-state component has the amplitude $\hat{I}_k$ and is called symmetrical short-circuit current. The value of the maximum asymmetrical short-circuit current $\hat{I}_{max}$ depends on phase angle ϑ and impedance X/R-ratio.
For usual power transformers and supply networks, the inductance outweighs the resistance $X_k \gg R_k$. It therefore follows $\varphi_k = \pi/2$. Two limiting cases can be considered for the the maximum short-circuit current:
a) The short-circuit happens during the voltage maximum ($\vartheta = \pi/2$). The maximum short-circuit current is equal to the steady-state short-circuit current $\hat{I}_k$:

$$\hat{I}_{max} = i_k(t)|_{t=1/(4f_0)} = \hat{I}_k \quad \text{for } \vartheta = \varphi = \pi/2 \tag{2.5}$$

b) If the short-circuit occurs during a voltage zero-crossing ($\vartheta = 0$), $\hat{I}_{max}$ reaches two times the steady state short-circuit current:

$$\hat{I}_{\max} = i_{\mathrm{k}}(t)\big|_{t=1/(2f_0)} = 2\hat{I}_{\mathrm{k}} \quad \text{for } \vartheta = 0 \text{ and } \varphi = \pi/2 \tag{2.6}$$

The factor of 2 between $\hat{I}_{\max}$ and $\hat{I}_{\mathrm{k}}$ is named asymmetry factor. Corresponding values for $1 < X/R \leq 100$ are given by [12, p. 17] in the form of a diagram.

2.2.2 Short-Circuit Forces

The fundamental Lorentz force law according to Equation (2.7) describes the total electromagnetic force $\mathbf{F}$ in Newton, experienced by a charge Q moving in a magnetic flux density $\mathbf{B}$ and an electrical field $\mathbf{E}$ with velocity $\mathbf{v}$. Bold variables refer to vectors.

$$\mathbf{F} = Q\left(\mathbf{E} + \mathbf{v} \times \mathbf{B}\right) \tag{2.7}$$

The first component $Q \cdot \mathbf{E}$ corresponds to the electric force and acts on both moving and stationary charges. The second component $Q \cdot \mathbf{v} \times \mathbf{B}$, however, only acts on moving charges. For the application of conductors exposed to magnetic fields the electrostatic contribution can be omitted [52].
Written in the means of space charge density ρ_{c} (Equation (2.8)) and current density $\mathbf{J}$ (Equation (2.9)), the Lorentz force equation simplifies to Equation (2.10). The differential $\mathrm{d}V$ represents an infinitesimal volume of a conductor.

$$Q = \int \rho_{\mathrm{c}} \mathrm{d}V \tag{2.8}$$

$$\mathbf{J} = \rho_{\mathrm{c}} \cdot \mathbf{v} \tag{2.9}$$

$$\mathbf{F} = \int \rho_{\mathrm{c}} \mathrm{d}V \cdot \mathbf{v} \times \mathbf{B} = \int \left\{\mathbf{J} \times \mathbf{B}\right\} \mathrm{d}V \tag{2.10}$$

In conclusion, electromagnetic forces arise in conductive parts carrying a current density $\mathbf{J}$ and being exposed to a magnetic flux density $\mathbf{B}$. Inside power transformers, this is the case for the coils and leads, which are exposed to the leakage flux in the core window as shown in Figure 2.6.

Regarding transformer windings it is usual to calculate the line load $\mathbf{q}$, which defines the force on a turn per unit length. Dividing Equation (2.10) by the conductor length L simplifies the volume integral $\mathrm{d}V$ to an integral over the conductor cross-sectional area $\mathrm{d}A$. Introducing the current $\mathbf{I} = \int \mathbf{J} \mathrm{d}A$ leads to:

$$\mathbf{q} = \frac{\mathbf{F}}{L} = \mathbf{I} \times \mathbf{B} = \begin{pmatrix} q_{\mathrm{r}} \\ q_{\varphi} \\ q_{\mathrm{z}} \end{pmatrix} = \begin{pmatrix} I_{\varphi} B_{\mathrm{z}} \\ 0 \\ -I_{\varphi} B_{\mathrm{r}} \end{pmatrix} \tag{2.11}$$

Inside concentric windings, the predominant current component flows in the azimuthal direction. Based on the cylindrical coordinate system of a coil, it is named I_φ in the following. As both radial and axial component I_r and I_φ are approximately zero, the azimuthal component B_φ for the leakage flux is zero. This splits the electromagnetic load into radial and axial components q_r and q_z, given by Equation (2.11).
Figure 2.6 visualizes the load direction inside a typical coil arrangement by the means of arrows. The load acting on the LV coil is radially inwards directed. The HV coil is stretched in the positive radial direction. Close to the top and bottom winding endings the leakage flux has a significant radial component. Therefore, axial compressive force components arise on the windings.

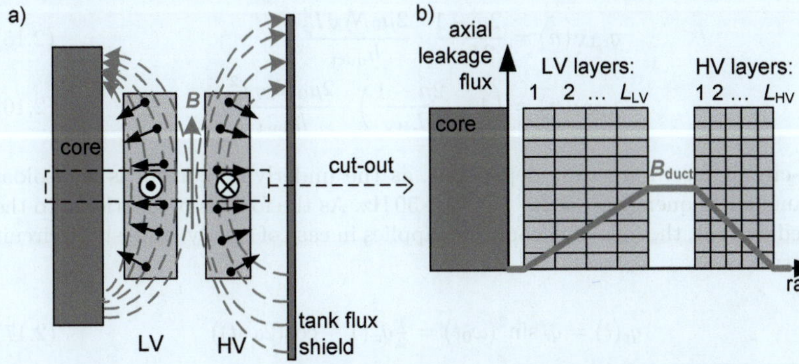

Figure 2.6: a) Leakage flux (green) and electromagnetic force distribution (black arrows) through LV, main duct and HV winding
b) Plot of the axial leakage flux component over radius

For most power transformers, the axial winding height is much larger than its width in the radial direction. Therefore, most of the coil length is exposed to an almost axially directed leakage flux B_z which is responsible for the radial short-circuit load q_r. A schematic plot of the flux B_z is shown in Figure 2.6 b). Beginning at the inner side of the LV winding it linearly rises from zero to the maximum value B_{duct}, which occurs in the main duct between LV and HV winding. B_{duct} can be approximated using Equation (2.12) with vacuum permeability μ_0, number of LV winding turns N_{LV}, duct height h_{duct} and winding current I_φ. Due to the ampere-turn balance between LV and HW coil the axial leakage flux drops from its maximum in the duct, to zero on the outer side of the HV winding.

$$B_{duct} = \sqrt{2}\mu_0 N_{LV} I_\varphi / h_{duct} \tag{2.12}$$

Assuming a LV winding with L_{LV} layers the average axial leakage flux through the middle of a winding in layer no. n is given by Equation (2.13). For a HV winding with L_{LV} layers Equation (2.14) is relevant. The layer index n increases from the inner to the outer radius, as shown in Figure 2.6.

$$\bar{B}_{\mathrm{z,LV}}(n) = \frac{2n-1}{2L_{\mathrm{LV}}} \cdot B_{\mathrm{duct}} \tag{2.13}$$

$$\bar{B}_{\mathrm{z,HV}}(n) = \left(1 - \frac{2n-1}{2L_{\mathrm{HV}}}\right) \cdot B_{\mathrm{duct}} \tag{2.14}$$

Finally, the product $\bar{B}_{\mathrm{z}}(n) \cdot \sqrt{2} I_{\varphi}$ results in the radial short-circuit load for each conductor inside a certain layer n of the LV or HW winding:

$$q_{\mathrm{r,LV}}(n) = \frac{2n-1}{2L_{\mathrm{LV}}} \cdot \frac{2\mu_0 N_{\mathrm{LV}} I_{\varphi}^2}{h_{\mathrm{duct}}} \tag{2.15}$$

$$q_{\mathrm{r,HV}}(n) = \left(1 - \frac{2n-1}{2L_{\mathrm{HV}}}\right) \cdot \frac{2\mu_0 N_{\mathrm{LV}} I_{\varphi}^2}{h_{\mathrm{duct}}} \tag{2.16}$$

Short-circuit forces are time dependent, as the underlying current is sinusoidal with angular frequency $\omega_0 = 2\pi \cdot f_0 = 2\pi \cdot 50\,\mathrm{Hz}$. As the force is proportional to the squared current, the following condition applies in case of steady-state short-circuit current:

$$q_{\mathrm{r}}(t) = \hat{q}_{\mathrm{r}} \sin^2(\omega_0 t) = \tfrac{1}{2}\hat{q}_{\mathrm{r}} \left(1 - \cos(2\omega_0 t)\right) \tag{2.17}$$

Therefore, the force is of unidirectional and $2f_0 = 100\,\mathrm{Hz}$ pulsating type. A separation into a constant part of $\hat{q}_{\mathrm{r}}/2$, and a time dependent part is possible. The index n of the layer is dropped for better clarity, and the amplitude is abbreviated by $\hat{q}_{\mathrm{r}}$.

2.2.3 Failures Modes

There are three groups of failures, depending on the force direction. Some of the failure modes are visualized in Figure 2.7.

Radial failures:

- Free buckling arises, if the stiffness of the winding's radial support structure is negligible and the spacers do not act as rigid support. Especially, if a tertiary winding between core and LV is present, the radial support structure for the LV winding is weak. Free buckling primarily is an issue for free supporting windings, which are designed in such a way to withstand radial compressive forces without the contribution of radial supports. In such a case, it is necessary to use epoxy-bonded CTCs or conductors made from high strength copper.
- Forced buckling is primary an issue if the spacers remain fixed, and withstand the radial forces. The winding cannot move in the radial direction, where it is in contact with the spacers. Due to the spacers, limiting the radial movement, it is possible to use conductors having lower buckling resistance, such as unbonded CTCs for example.
- Radial forces on the HV side stretch the coil radially outwards. This load direction results in a less critical hoop stress compared to the radial compressive forces. At the most this can lead to tearing of soldered joints.

Axial failures:

- Tilting of conductors arises due to axial compressive forces on the coils. If these forces exceed a certain limit, windings can fail in a zig-zag pattern. The tilting strength mainly is dependent on the conductor dimensions. Twisting strength increases with conductor thickness in the radial direction. Thus, the risk of tilting increases for thin and high conductors.
- A decrease in coil clamping forces is an issue if insulating materials like Pressboard spacers and conductor insulation are compressed plastically. In turn, oscillatory forces acting on the insulation increase. Also, mechanical resonance frequencies of the winding decrease.
- Axial bending between radial spacers can occur especially for conductors with a small height. Conductor bending can lead to insulation damage.
- Thrust on a whole coil arises from axially unbalanced coils from the same limb. The resulting axial forces can displace either a complete coil, or at least some layers. The collapse of the axial supporting structure is another possible failure.

Combined failures:

- Spiraling is the most common combined failure in LV helical windings. The helix angle, radial and axial forces and the end sections influence the spiraling force, which has the tendency to tighten the LV coil. These forces mainly occur at the end turns, where the radial flux is high.

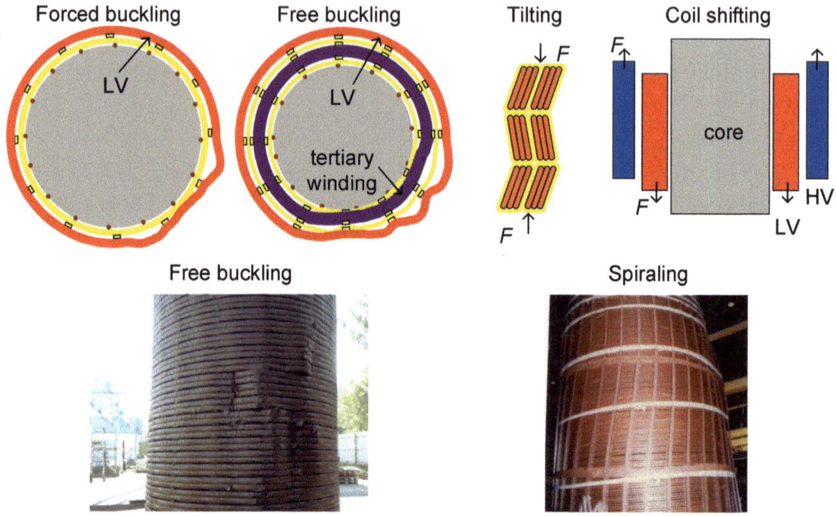

Figure 2.7: Effects of radial, axial and combined short-circuit forces (spiraling picture taken from [22])

2.3 Mechanics of Materials

2.3.1 Elastic Behavior of Materials

One of the most important mechanical parameter of materials is the Young's modulus E_0. It defines the correlation between normal stress σ and normal strain ϵ for small loads. The normal stress is defined according to Equation (2.18), with force F and cross-section A. The normal average strain can be written in the means of Equation (2.19), with origin member length L and elongation ΔL (compare Figure 2.8 a)).

$$\sigma = F/A \tag{2.18}$$

$$\epsilon = \Delta L/L \tag{2.19}$$

As long as a material behaves linear elastic, Hooke's law is valid:

$$\sigma = \epsilon E_0 \tag{2.20}$$

However, a tensile strain (positive sign) in one direction of a member will almost always result in a compressive strain (negative sign) in the other direction. For example in Figure 2.8 b) the cube contracts perpendicular to the load direction.

The underlying material parameter describing this behavior is the Poisson's ratio ν_p. It relates lateral strain ϵ_{lat} and longitudinal strain ϵ_{long} according to

$$\nu_p = \frac{-\epsilon_{lat}}{\epsilon_{long}}. \tag{2.21}$$

The Poisson's ratio lies between $0 \leq \nu_p < 0.5$ for common materials. Only special materials like composites, foams or auxetic polymers may differ from this behavior with for example $\nu_p < 0$.

Copper alloys have a Poisson's ratio of 0.3 [3], which is valid for almost every metal. However, a precise measurement of ν_p is not crucial, as it usually does not strongly affect the results of structural calculations or simulations, in contrast to the Young's modulus.

Apart from elongational or compressive forces the deformation of structures can also include shearing loads, as shown in Figure 2.8 c). The shearing stress τ_s can be expressed in the means of shearing force F and cross-sectional area A according to

$$\tau_s = F/A. \tag{2.22}$$

The shear modulus G describes the linear relationship between shearing stress τ_s and shearing angle γ_s:

$$\tau_s = G\gamma_s \tag{2.23}$$

The shear modulus of any material has to be between $\frac{1}{3}E_0$ and $\frac{1}{2}E_0$. This results from the fact that all three parameters ν_p, E_0 and G are not independent for a given material [11]. They have to satisfy the relationship

$$\frac{E_0}{2G} = 1 + \nu_p. \tag{2.24}$$

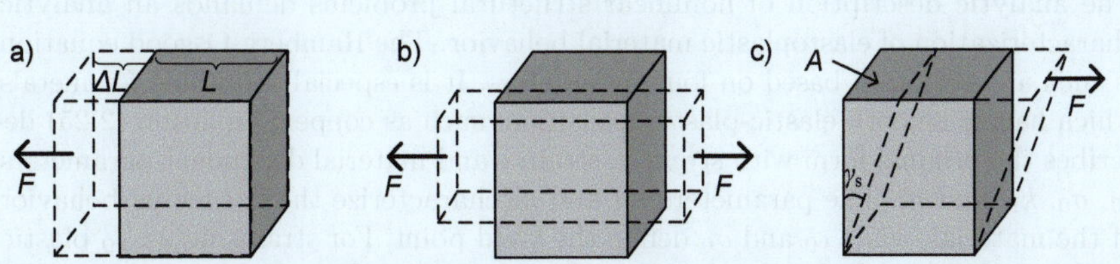

Figure 2.8: a) Member with an axial load, b) member under longitudinal tensile load showing lateral compression, c) member with shearing load

2.3.2 Theory of Plasticity

Both elastic and plastic deformation of materials can be explained on the atomic scale. Each deformation causes a change in material volume, and a shift in shape at constant volume. The latter is known as shear deformation. If this process causes atoms to move to new positions, which preserve the configuration of the crystal lattice, plastic deformation occurs. The part of shear which recovers corresponds to elastic deformation.
If complete planes of atoms slip, the boundary of the slipped area is called dislocation. In ductile metals like copper, for example, dislocations are responsible for plastic deformation. The energy which is necessary to move dislocations is much lower than the energy which would lead to fracture.
Dislocations control the yield strength and hardening characteristics of ductile metals. Obstacles like grain boundaries, impurity atoms or intersecting dislocation boundaries prevent the movement of dislocations, resulting in a higher yield strength.
A commonly used forming process is cold-working, which is done at temperatures far below the recrystallization temperature of the metal. During this process, new dislocations and intersecting boundaries between dislocations develop. These obstacles increase the resistance against further movements of dislocations, resulting in a hardening of the metal with proceeding deformation. The underlying phenomenon is also named strain-hardening, as the strain fields of adjacent dislocations increase the withstand capability to further dislocation movement. Furthermore, the distorted grain structure is also responsible for an increase in yield strength, as the impeded dislocation movement will hinder the onset of plasticity.
Annealing retracts these effects of strain hardening. This heat treatment reduces the accumulation of dislocations and retains the grain structure of the metal.

2.3.3 Ramberg-Osgood Equation

The analytic description of nonlinear structural problems demands an analytic characterization of elastoplastic material behavior. The Ramberg-Osgood equation is such a description based on four parameters. It is especially suitable for metals which have a smooth elastic-plastic transition, such as copper. Equation (2.25) describes the original form with stress σ, strain ϵ and material dependent parameters ϵ_0, σ_0, k_R and m. The parameters k_R and m characterize the hardening behavior of the material, while ϵ_0 and σ_0 define the yield point. For strains above ϵ_0 plastic strain predominates the elastic strain. The Ramberg-Osgood equation provides close approximations for stress-strain measurements up to the 0.2 % yield stress [56].

$$\frac{\epsilon}{\epsilon_0} = \frac{\sigma}{\sigma_0} + k_\mathrm{R}\left(\frac{\sigma}{\sigma_0}\right)^m \tag{2.25}$$

The tangent modulus $E_t(\sigma) = \mathrm{d}\sigma/\mathrm{d}\epsilon$ describes the slope of the stress-strain characteristic $\sigma(\epsilon)$. This function $E_t(\sigma)$ often is used as stress dependent Young's modulus in analytic calculations. However, solving Equation (2.25) for σ is not possible. Solving Equation (2.25) for ϵ and introducing the linear Young's modulus $E_0 = \sigma_0/\epsilon_0$ leads to

$$\epsilon(\sigma) = \frac{\sigma}{E_0} + k_\mathrm{R}\left(\frac{\sigma}{\sigma_0}\right)^m \frac{\sigma_0}{E_0}. \tag{2.26}$$

Using the inverted derivation of Equation (2.26) gives

$$\begin{aligned} E_t(\sigma) = \frac{\mathrm{d}\sigma}{\mathrm{d}\epsilon} = \left(\frac{\mathrm{d}\epsilon}{\mathrm{d}\sigma}\right)^{-1} &= \left(\frac{1}{E_0} + \frac{k_\mathrm{R} m}{E_0}\left(\frac{\sigma}{\sigma_0}\right)^{m-1}\right)^{-1} \\ &= \frac{E_0}{1 + k_\mathrm{R} m \left(\frac{\sigma}{\sigma_0}\right)^{m-1}}. \end{aligned} \tag{2.27}$$

[60] proposes $k_\mathrm{R} = 3/7$ for the characterization of copper alloys. A curve fit of the Ramberg-Osgood equation to a stress-strain characteristic deduced from tensile tests reveals the remaining parameters.

The yield point (also known as proof stress) characterizes the lowest stress at which plastic deformation occurs. This value is often used to distinguish between different alloys of a specific metal.

Copper alloys do not exhibit a distinct yield point, therefore the so-called offset yield point is an alternative parameter. Usually, the offset yield point refers to 0.2 % plastic strain. The corresponding stress is named $\sigma_{0.2}$. In many material datasheets this value is also known as $R_{\mathrm{p},0.2}$. The underlying definition is the intersection point between the nonlinear stress-strain relationship and a line starting at $\epsilon = 0.2\,\%$ with a gradient equal to the linear Young's modulus E_0, as shown in Figure 2.9 a). This calculation leads to

$$R_{\mathrm{p}0.2} = \sigma_{0.2} = \sigma_0^{\frac{m-1}{m}} \left(\frac{E_0 \epsilon}{k_\mathrm{R}}\right)^{\frac{1}{m}} \Bigg|_{\epsilon = 0.2\,\%}. \tag{2.28}$$

For strains above $\epsilon_{0.2}$, the Ramberg-Osgood approach may become inaccurate. Even if the curve fit data range is increased to the full-range stress-strain curve, the Ramberg-Osgood model does not match the measurement data for strains below $\epsilon_{0.2}$.

However, the author of [56] presented an extended Ramberg-Osgood model for stainless steel alloys. The subsequent steps explain how this approach can be adapted to copper stress-strain characteristics:

The measurement data range has to be divided into two parts:

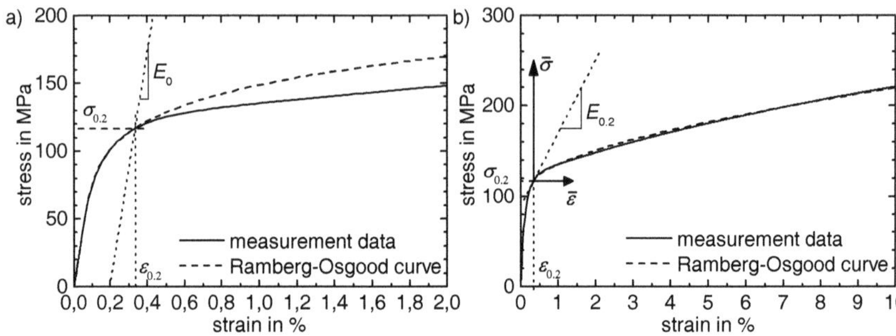

Figure 2.9: a) Stress-strain data with Ramberg-Osgood fit for $0 < \epsilon \leq \epsilon_{0.2}$
b) Stress-strain data with Ramberg-Osgood fit for $\epsilon > \epsilon_{0.2}$

Between $0 < \epsilon \leq \epsilon_{0.2}$ a nonlinear curve fit based on Equation (2.26) with fixed $k_{\mathrm{R}} = 3/7$ estimates the basic Ramberg-Osgood parameters.
An extended Ramberg-Osgood equation with transformed stress and strain according to

$$\bar{\epsilon} = \epsilon - \epsilon_{0.2} \tag{2.29}$$

$$\bar{\sigma} = \sigma - \sigma_{0.2} \tag{2.30}$$

describes the measurement data for $\epsilon > \epsilon_{0.2}$. The transformation is visualized in Figure 2.9 b) and leads to the following equation:

$$\bar{\epsilon}(\bar{\sigma}) = \frac{\bar{\sigma}}{E_{0.2}} + k_{\mathrm{R}} \left(\frac{\bar{\sigma}}{\tilde{\sigma}_0}\right)^{\tilde{m}} \frac{\tilde{\sigma}_0}{E_{0.2}} \tag{2.31}$$

The new parameters $\tilde{m}$ and $\tilde{\sigma}_0$ are marked by a tilde. Factor k_{R} remains at its original value of $3/7$.
To fulfill steadiness in the point $(\epsilon_{0.2}, \sigma_{0.2})$, the initial modulus $E_{0.2}$ given in Equation (2.31) has to correspond to the tangent modulus of Equation (2.26) in $\epsilon = \epsilon_{0.2}$. The Young's modulus of the extended Ramberg-Osgood equation is

$$E_{0.2} = \frac{E_0}{1 + \frac{3}{7}\tilde{m}\left(\frac{\sigma_{0.2}}{\sigma_0}\right)^{\tilde{m}-1}}. \tag{2.32}$$

Finally, the full-range stress-strain curve in inverse representation is given by

$$\epsilon(\sigma) = \begin{cases} \dfrac{\sigma}{E_0} + \dfrac{3}{7}\left(\dfrac{\sigma}{\sigma_0}\right)^m \dfrac{\sigma_0}{E_0} & \text{for} \quad \sigma \leq \sigma_{0.2} \\ \dfrac{\sigma - \sigma_{0.2}}{E_{0.2}} + \dfrac{3}{7}\left(\dfrac{\sigma - \sigma_{0.2}}{\tilde{\sigma}_0}\right)^{\tilde{m}} \dfrac{\tilde{\sigma}_0}{E_{0.2}} + \epsilon_{0.2} & \text{for} \quad \sigma > \sigma_{0.2}. \end{cases} \tag{2.33}$$

The full-range tangent modulus results in

$$E_t(\sigma) = \begin{cases} \dfrac{E_0}{1+\frac{3}{7}m\left(\frac{\sigma}{\sigma_0}\right)^{m-1}} & \text{for} \quad \sigma \leq \sigma_{0.2} \\ \dfrac{E_{0.2}}{1+\frac{3}{7}\tilde{m}\left(\frac{\sigma-\sigma_{0.2}}{\tilde{\sigma}_0}\right)^{\tilde{m}-1}} & \text{for} \quad \sigma > \sigma_{0.2}. \end{cases} \tag{2.34}$$

Two independent curve fit processes estimate the following set of 5 parameters: E_0, m, $\tilde{m}$, σ_0 and $\tilde{\sigma}_0$.
Figure 2.10 presents a full-range Ramberg-Osgood approximation for the exemplary measurement data according to Figure 4.4. The strain axis is scaled logarithmically for better clearness and to show the close approximation of the measurement data.

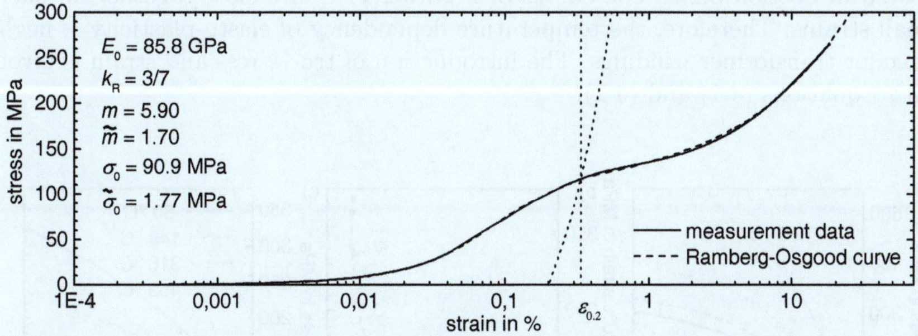

Figure 2.10: Stress-strain data with extended Ramberg-Osgood curve fit

2.3.4 Strain Rate and Temperature Dependency of Copper

The mechanical properties of metals depend on the temperature and the strain rate caused by the loading. The elastoplastic properties are most affected, due to the dislocations, whose movement is a temperature dependent process. With increasing temperature the thermal atomic vibration simplifies the bypassing of obstacles for the dislocation lines, thus reducing the yield strength [30]. An increase in strain rate reduces the time which is available for thermal activation. In conclusion, the yield strength increases again. According to [30], a strain rate increase is equivalent to a decrease in temperature, as a good first approximation.
The authors of [29] provided a summarizing literature review with focus on the mechanical properties of OFHC copper for strain rates ranging from $10^{-3} \ldots 10^{5}\ \mathrm{s}^{-1}$. Figure 2.11 a) provides some sample stress-strain curves. This data shows the yield strength increase with strain rate increase. Plotted in the means of stress at 20 % strain over the strain rate leads to Figure 2.11 b). The main result is that the OFHC copper yield strength increases slightly for strain rates up to $10^{3}\ \mathrm{s}^{-1}$. Higher rates however, lead to a disproportionately high increase.

Section 3.2.2 provides a brief analysis of typical strain rates occurring inside power transformer windings: For the elastic 100 Hz vibration during normal operation the strain rate stays far below $1\,s^{-1}$. Higher values up to $5\,s^{-1}$ only occur in the end phase of plastic buckling caused by excessive electromagnetic short-circuit loads.

In oil-paper insulated power transformers the operating temperature close to the copper wires usually stays below 90 °C. The temperature dependency of the copper's Young's modulus is almost negligible for typical winding temperatures, as the summarizing Table 2.1 taken from [61] proves. The difference between 20 °C and 100 °C is below 1 % as long as the cold working part does not exceed 20 %. In contrast, the change in elastoplastic material behavior is more temperature dependent: Figure 2.11 c) visualizes the plastic part of true stress-strain characteristics of OFHC copper for temperatures between −50 °C up to 482 °C. However, strains arising during the buckling of transformer windings are below 0.01, as proved in Section 3.2.2. The curves of Figure 2.11 c) are close together for such small strains. Therefore, the temperature dependency of elasto-plasticity is negligible for transformer windings. The introduction of true stress and strain is given with Equations (4.1) and (4.2).

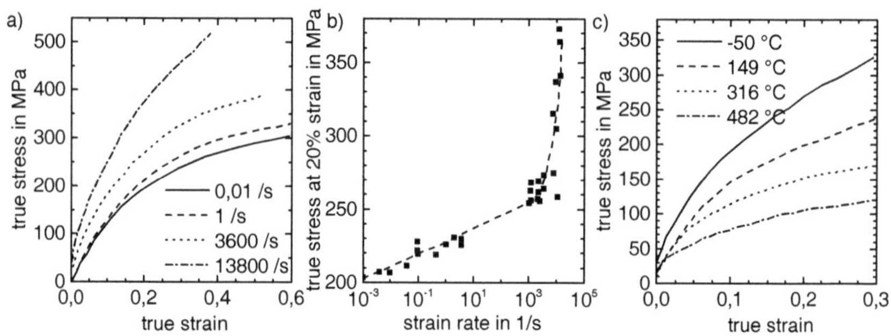

Figure 2.11: a) OFHC copper strain rate dependency (data: [29])
b) OFHC copper true stress at 20 % strain (data: [29])
c) OFHC copper temperature dependency (data: [4])

Table 2.1: Copper Young's modulus temperature dependency (data: [61])

	annealed	20 % cold worked	30 % cold worked
20 °C	107 GPa	110 GPa	127 GPa
100 °C	106 GPa	111 GPa	115 GPa
200 °C	98 GPa	104 GPa	110 GPa

2.4 Method of Finite Element Analysis

The Finite Element Analysis (FEA) provides a powerful calculation method for the approximate solution of a wide range of engineering problems, such as electrical, mechanical, thermal or fluid based tasks, or a combination of some in a multiphysics approach. The problem, which shall be solved, has to be a limited continuum (e.g. a solid). Field variables characterize the solution of the investigated physical phenomenon. Some examples are displacements in solid mechanics, electric and magnetic potential in electrical or temperature fields in thermal problems.
The general procedure is to decompose a given domain, the continuum, into a finite number of subdomains (elements). The unknown field variables inside the domain are expressed in terms of interpolation functions within each element. These functions are defined by the coordinates of the element nodes and the unknown values of the field variables at these nodes. Finally, if a solution for the nodal field is present, interpolation functions allow evaluating the field variables at any point in the element.
Each subdomain needs a so-called element property matrix. The most common approaches are:

- Direct approach: A direct relationship between fundamental physics and nature of the problem is given.
- Variational or energy approach: A variational functional (e.g. potential energy in structural mechanics problems) describes the problem.
- Weighted residuals: The element matrices directly result from the governing equations of the problem.

For more details to these approaches refer to [14, 55, 41].

The assembly of all element matrices results in the final global system matrix $\mathbf{K}$. The final equation system is solvable after the insertion of boundary conditions (e.g. a prescribed displacement). Written in matrix form gives

$$\mathbf{K} \cdot \mathbf{u} = \mathbf{F} \tag{2.35}$$

with unknown field variables $\mathbf{u}$ of the nodal points and exciting variables $\mathbf{F}$. The dimension of the global system matrix $\mathbf{K}$ is equal to the number of global nodes times number of unknowns (degree of freedoms) per node. For linear analysis ($\mathbf{K}$ is independent of $\mathbf{u}$) Equation (2.35) is solved for $\mathbf{u}$ by inversion, as given in Equation (2.36). Depending on the dimension of $\mathbf{K}$ either a direct or iterative solver is necessary.

$$\mathbf{u} = \mathbf{K}^{-1} \cdot \mathbf{F} \tag{2.36}$$

The field variables at any point within the domain result from the final solution $\mathbf{u}$ by using interpolation functions.

2.4.1 Magnetic Formulation

The basic equations of magnetostatic fields are Ampere's Law (Equation (2.37)), the conservation of magnetic flux $\mathbf{B}$ (Equation (2.38)) and the constitutive relation according to Equation (2.39) in terms of magnetic field $\mathbf{H}$, external current density $\mathbf{J}$ and permeability μ.

$$\nabla \times \mathbf{H} = \mathbf{J} \tag{2.37}$$
$$\nabla \cdot \mathbf{B} = 0 \tag{2.38}$$
$$\mathbf{B} = \mu \mathbf{H} \tag{2.39}$$

Substituting the magnetic vector potential $\mathbf{A}$ according to $\mathbf{B} = \nabla \times \mathbf{A}$ into Equation (2.39) leads to

$$\mathbf{H} = \frac{1}{\mu} \nabla \times \mathbf{A}. \tag{2.40}$$

With this equation the replacement of $\mathbf{H}$ in Equation (2.37) results in $\nabla \times \nabla \times \mathbf{A} = \mu \mathbf{J}$. Assuming the Coulomb Gauge condition $\nabla \cdot \mathbf{A} = 0$ transfers this expression into Poisson's equation for magnetostatic fields:

$$\nabla^2 \mathbf{A} = -\mu \mathbf{J} \tag{2.41}$$

In the 2D case the FEA formulation of Equation (2.41) is given in matrix form $\mathbf{K}_{\mathrm{e,m}} \cdot \mathbf{A} = \mathbf{F}_{\mathrm{e,m}}$ with element magnetic stiffness matrix $\mathbf{K}_{\mathrm{e,m}}$ and element magnetic force vector $\mathbf{F}_{\mathrm{e,m}}$.
According to [16, p. 33] the entries of $\mathbf{K}_{\mathrm{e,m}}$ and $\mathbf{F}_{\mathrm{e,m}}$ are

$$k_{\mathrm{e,m},jk} = -\int_x \int_y \frac{1}{\mu} \left(\frac{\partial N_j}{\partial x} \frac{\partial N_k}{\partial x} + \frac{\partial N_j}{\partial y} \frac{\partial N_k}{\partial y} \right) \mathrm{d}x \mathrm{d}y \tag{2.42}$$
$$f_{\mathrm{e,m},jk} = \int_x \int_y \left(J_z N_j \right) \mathrm{d}x \mathrm{d}y. \tag{2.43}$$

with the indices j, k corresponding to the numbering of the finite element nodes. N_j describes the interpolation function of finite element node no. j.
After element assembly, the governing matrix equation can be written in the form of Equation (2.35).

2.4.2 Structural Mechanics Formulation

This section presents a brief derivation of the governing equations used in a structural mechanics FEA formulation, summarized from [41, p. 222].
Generally, in a structural mechanics problem, the following quantities describe the behavior of a structure subjected to deforming influences: Displacements u_j, normal stresses σ_j, shear stresses τ_{jk}, normal strains ϵ_j and shearing strains γ_{jk}, with $j, k =$ x,y,z being the Cartesian coordinates. The representation in vector formulation is:

$$\begin{aligned} \mathbf{u}^\mathsf{T} &= \{u_\mathrm{x}\ u_\mathrm{y}\ u_\mathrm{z}\} \\ \boldsymbol{\sigma}^\mathsf{T} &= \{\sigma_\mathrm{x}\ \sigma_\mathrm{y}\ \sigma_\mathrm{z}\ \tau_\mathrm{xy}\ \tau_\mathrm{yz}\ \tau_\mathrm{zx}\} \\ \boldsymbol{\epsilon}^\mathsf{T} &= \{\epsilon_\mathrm{x}\ \epsilon_\mathrm{y}\ \epsilon_\mathrm{z}\ \gamma_\mathrm{xy}\ \gamma_\mathrm{yz}\ \gamma_\mathrm{zx}\} \end{aligned} \tag{2.44}$$

The property matrix $\mathbf{C}$ (also known as elasticity matrix) relates stresses and strains by the means of Young's modulus E_0 and Poisson's ratio ν_p for an isotropic, linear elastic material:

$$\boldsymbol{\sigma} = \frac{E_0}{(1+\nu_\mathrm{p})(1-2\nu_\mathrm{p})} \begin{Bmatrix} 1-\nu_\mathrm{p} & \nu_\mathrm{p} & \nu_\mathrm{p} & 0 & 0 & 0 \\ \nu_\mathrm{p} & 1-\nu_\mathrm{p} & \nu_\mathrm{p} & 0 & 0 & 0 \\ \nu_\mathrm{p} & \nu_\mathrm{p} & 1-\nu_\mathrm{p} & 0 & 0 & 0 \\ 0 & 0 & 0 & \frac{1-2\nu_\mathrm{p}}{2} & 0 & 0 \\ 0 & 0 & 0 & 0 & \frac{1-2\nu_\mathrm{p}}{2} & 0 \\ 0 & 0 & 0 & 0 & 0 & \frac{1-2\nu_\mathrm{p}}{2} \end{Bmatrix} \boldsymbol{\epsilon} \tag{2.45}$$

Expressed in the means of the displacement components $\mathbf{u}$, the strain vector $\boldsymbol{\epsilon}$ is given by

$$\boldsymbol{\epsilon} = \begin{Bmatrix} \partial/\partial x & 0 & 0 \\ 0 & \partial/\partial y & 0 \\ 0 & 0 & \partial/\partial z \\ \partial/\partial y & \partial/\partial x & 0 \\ 0 & \partial/\partial z & \partial/\partial y \\ \partial/\partial z & 0 & \partial/\partial x \end{Bmatrix} \mathbf{u} \quad \text{or} \quad \boldsymbol{\epsilon} = \mathbf{L}\mathbf{u} \tag{2.46}$$

with differential-operator matrix $\mathbf{L}$.
The final governing equations result from a variational approach: If the total potential energy $\Pi = \sum \Pi_e$ gets minimal, the finite element method has obtained the best approximation for the field variables. The potential energy Π_e for an element with the number e is

$$\Pi_e = \frac{1}{2}\int_{\Omega_e} \boldsymbol{\sigma}^\mathsf{T}\boldsymbol{\epsilon}\mathrm{d}V_e - \int_{\Omega_e} \mathbf{u}^\mathsf{T}\mathbf{f}_\mathrm{v}\mathrm{d}V_e - \int_{\Gamma_e} \mathbf{u}^\mathsf{T}\mathbf{f}_\mathrm{s}\mathrm{d}S_e \tag{2.47}$$

with volume, respective surface force vectors $\mathbf{f}_\mathrm{v}$ and $\mathbf{f}_\mathrm{s}$. The volume and surface integrals are defined over the element volume Ω_e and its boundary surface Γ_e. The last two integrals correspond to the potential energy caused by volume, respective surface loads. The first term in the definition of Π_e represents the internal strain energy.
The correlation between approximated displacements $\tilde{\mathbf{u}}$ and nodal displacements $\mathbf{u}_\mathrm{n}$ is given by the matrix of the interpolation functions $\mathbf{N}$ to $\tilde{\mathbf{u}} = \mathbf{N}\mathbf{u}_\mathrm{n}$. Using this approximated form and Equation (2.46) calculates the strain components within each element to

$$\boldsymbol{\epsilon} \approx \mathbf{L}\tilde{\mathbf{u}} = \mathbf{L}\mathbf{N}^\mathsf{T}\mathbf{u}_\mathrm{n} = \mathbf{P}\mathbf{u}_\mathrm{n}, \tag{2.48}$$

with strain-displacement matrix $\mathbf{P}$.
The correspondence $\boldsymbol{\sigma} = \mathbf{C}\boldsymbol{\epsilon}$ links stresses and strains based on the elasticity matrix $\mathbf{C}$ according to Equation (2.45).
Introducing $\mathbf{P}$, $\mathbf{N}$ and $\mathbf{C}$ into Equation (2.47) leads to the following expression for the total potential energy in the means of element nodal displacements $\mathbf{u}_\mathrm{n}$:

$$\Pi_\mathrm{e} = \frac{1}{2}\int_{\Omega_\mathrm{e}} \mathbf{u}_\mathrm{n}^\mathsf{T}\left(\mathbf{P}^\mathsf{T}\mathbf{C}\mathbf{P}\right)^\mathsf{T}\mathbf{u}_\mathrm{n}\mathrm{d}V - \int_{\Omega_\mathrm{e}} \mathbf{u}_\mathrm{n}^\mathsf{T}\mathbf{N}^\mathsf{T}\mathbf{f}_\mathrm{v}\mathrm{d}V - \int_{\Gamma_\mathrm{e}} \mathbf{u}_\mathrm{n}^\mathsf{T}\mathbf{N}^\mathsf{T}\mathbf{f}_\mathrm{s}\mathrm{d}S \tag{2.49}$$

The minimization of potential inside one element requires that

$$\frac{\partial \Pi_\mathrm{e}}{\partial \mathbf{u}_\mathrm{n}} = \frac{1}{2}\int_{\Omega_\mathrm{e}} \left(\mathbf{P}^\mathsf{T}\mathbf{C}\mathbf{P}\right)^\mathsf{T}\mathbf{u}_\mathrm{n}\mathrm{d}V - \int_{\Omega_\mathrm{e}} \mathbf{N}^\mathsf{T}\mathbf{f}_\mathrm{v}\mathrm{d}V - \int_{\Gamma_\mathrm{e}} \mathbf{N}^\mathsf{T}\mathbf{f}_\mathrm{s}\mathrm{d}S = 0. \tag{2.50}$$

Equation (2.50) is the element equilibrium equation in the form of $\mathbf{K}_\mathrm{e}\mathbf{u}_\mathrm{n} = \mathbf{F}_\mathrm{e}$ with element stiffness matrix $\mathbf{K}_\mathrm{e} = \int_{\Omega_\mathrm{e}}\left(\mathbf{P}^\mathsf{T}\mathbf{C}\mathbf{P}\right)^\mathsf{T}\mathrm{d}V$ and the element load vector $\mathbf{F}_\mathrm{e} = \int_{\Omega_\mathrm{e}}\mathbf{N}^\mathsf{T}\mathbf{f}_\mathrm{v}\mathrm{d}V + \int_{\Gamma_\mathrm{e}}\mathbf{N}^\mathsf{T}\mathbf{f}_\mathrm{s}\mathrm{d}S$.
The final system matrix $\mathbf{K}$ and the global nodal load vector $\mathbf{F}$ both assemble from the element values $\mathbf{K}_\mathrm{e}$, $\mathbf{F}_\mathrm{e}$. This leads to the final representation given in Equation (2.35). Insertion of boundary constraints, which prevent rigid-body motion, make this system solvable. Otherwise $\mathbf{K}$ becomes singular.

2.4.3 Coupling of the Magnetic and Mechanical Field

One conclusion of Section 2.2.2 has been, that magnetic forces result from the interaction between the stray magnetic field and the current flowing inside conductors. Neglecting the volume integral of Equation (2.10) leads to the volume correlated magnetic force density $\mathbf{f}_\mathrm{v}$ according to

$$\mathbf{f}_\mathrm{v} = \mathbf{J} \times \mathbf{B} = \mathbf{J} \times \left(\nabla \times \mathbf{A}\right). \tag{2.51}$$

A coupling to structural mechanics formulation is given if the magnetic force density replaces the external force in Equation (2.47).

2.4.4 Eigenvalue Buckling Analysis

The eigenvalue buckling analysis estimates buckling loads for an ideal elastic structure without imperfections. This kind of stability analysis consists of two steps: A small pre-load F_p applied to the structure, causes a linear elastic pre-buckling state. In this state, the so-called stress-stiffening effect (expressed by the stress stiffness matrix $\mathbf{S}$) modifies the structural stiffness (described by the stiffness matrix $\mathbf{K}$). $\mathbf{S}$ contributes to the stiffness matrix.
A subsequent eigenvalue analysis calculates numerous buckling modes which lead to instability. Instability occurs, if a structure undergoes great displacements for small load increments. The FEA solver has to solve the following equation for linear elastic eigenvalue buckling, with λ_j being the pre-load multiplier for the j-th eigenvalue:

$$(\mathbf{K} + \lambda_j \mathbf{S})\,\boldsymbol{\psi} = 0 \tag{2.52}$$

The vector $\boldsymbol{\psi}$ is the eigenvector corresponding to the eigenvalue. Usually, not all eigenvalues need to be determined. Only a small number of the lowest eigenvalues often is sufficient, as the lowest critical load is of main interest.
The critical buckling force depends on the load multiplier by

$$F_\mathrm{cr} = \lambda_j F_\mathrm{p}. \tag{2.53}$$

A nonlinear buckling analysis takes the variation of stiffness matrix $\mathbf{K}$ into account. These nonlinearities can be material plasticity, nonlinear contacts or large deformations, just to name a few. The underlying approach is to substitute $\mathbf{K}$ in Equation (2.52) with the tangent stiffness matrix $\mathbf{K}^\mathsf{T}$.

Chapter 3

Buckling Analysis

The electromagnetic forces are of oscillatory nature, as derived in Section 2.2.2. They act on the elastic system consisting of winding, insulation system and the support structure. These forces have to stay beyond a certain limit, to prevent permanent deformations of the coils with the consequences summarized in Section 2.2.3. Furthermore, resonance frequencies which could be excited by the time-dependent short-circuit current have to be much higher than the excitatory frequency of 100 Hz. Otherwise, small and elastic winding deformations might increase considerably by resonance effects.
Especially, the axial force components require a dynamic analysis. The natural frequencies in this direction might be close to the excitation frequency due to the large height of a coil and the easily compressible insulating materials. The large number of conductors in the axial direction, separated by insulating materials, acts like an elastic column with distributed mass and spring parameters [36, p. 247].
Concerning the radial force direction, a dynamic analysis usually is not an issue. The conductors have a high stiffness and a small mass, resulting in natural frequencies much higher than 100 Hz [36, p. 247]. In conclusion, the dynamic radial forces usually are considered slow, and the maximum stress is correlated to the maximum short-circuit current [43].
Based on this assumption, the present chapter presents fundamental equations for the static radial buckling analysis. A more detailed dynamic approach investigates hydrodynamic, inertial effects of the surrounding oil.

The theory of circular arches provides the fundamental approaches to analyze radial buckling: A transformer winding supported by spacers, subjected to an electromagnetic force, is quite similar to a circular arch under a radial compressive load. One critical assumption is, that the radial electromagnetic force is almost uniformly distributed within the middle-height section of a coil. This indicates that each winding out of this section is loaded uniformly, and it is sufficient to regard only a winding section.

The most simple case is the free buckling phenomenon, where the contribution of the radial support structure is negligible. The fundamental stability equation for an unsupported circular ring according to [17, p. 299] calculates the lowest critical load to

$$q_{\mathrm{cr}} = 3E_0 I/R^3. \tag{3.1}$$

E_0 is the conductor Young's modulus, R the mean winding radius and I the conductor area moment of inertia according to Equation (3.2) for bending around the coils principle z-axis. Section 3.1.1 provides more details to the area moment of inertia.

The analysis of forced buckling phenomena is more complicated than the free buckling, due to the constraints which are given by the inner support structure of a winding. Therefore, the present chapter focuses on this kind of failure.
The general approach regarding forced buckling is to consider a winding section between three spacers (Figure 3.1), as presented in previous works of [54, 60, 57]. Depending on the support antisymmetric or symmetric failure patterns arise:

- In the antisymmetric case, as visualized in Figure 3.1 a), one winding segment deforms radially inwards, while the other moves outwards. This effect benefits from small, weak spacers, which facilitate a rotation of the conductor at the spacer positions. Assuming such hinged supports the bending moment inside the conductor is almost zero at these hinges in the case of antisymmetric buckling shape.
- The symmetric buckling mode is shown in part b) of Figure 3.1. The spacers are rigid and vast and prevent a rotation of the conductor cross-section at their middle section. This concludes that each sector will deform independently from the other, leading to a symmetric buckling mode.

The underlying beam model (Euler-Bernoulli or Timoshenko) used for the buckling analysis depends on the ratio of conductor thickness in the radial direction and span between two supports. It is accepted, that if this thickness to span ratio exceeds 10 %, the Timoshenko beam model should be used, as it includes shear deformation effects [42]. In the case of power transformers, this only might be an issue for epoxy bonded CTCs with large radial widths. For unbonded CTCs consisting of thin, unbonded strands or for flat conductors, the Euler-Bernoulli model of slender beams is sufficient.

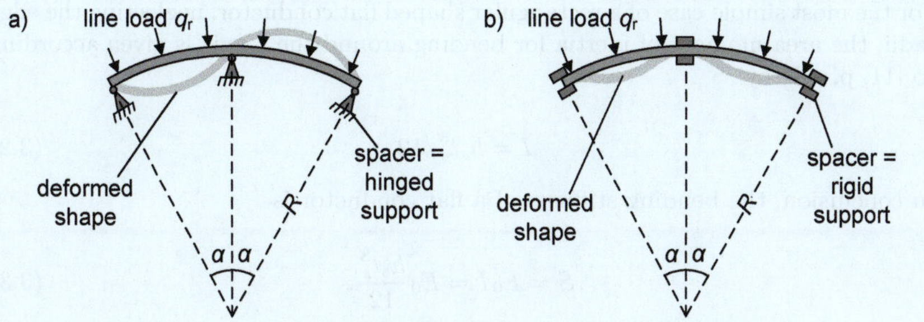

Figure 3.1: a) Antisymmetric radial buckling with weak spacers
b) Symmetric radial buckling with rigid spacers

3.1 Analytical Approach

3.1.1 Bending Stiffness of Conductors

The bending behavior of conductors is a fundamental property, which is necessary for buckling analysis of transformer windings. The bending stiffness as an underlying parameter is defined by the product of Young's modulus E_0 as material property and area moment of inertia I as geometry property. This product already occurred in Equation (3.1), describing the critical load of free buckling phenomena. The area moment of inertia is dependent on the axis of bending. In the present case of radially directed forces, the windings are bent in the radial direction concerning the coil's z-axis, as it is visualized in Figure 3.2. In this graph, the conductor orientation with respect to the coil's axial and radial axis corresponds to the common orientation used in power transformers. The dashed lines visualize the neutral fiber, where the bending stress is zero. More details regarding the mathematical background of this section are given in Appendix A.1.

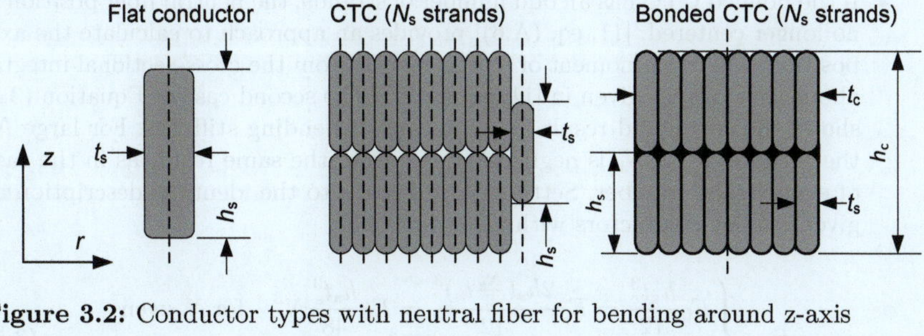

Figure 3.2: Conductor types with neutral fiber for bending around z-axis

For the most simple case of a rectangular shaped flat conductor, neglecting the edge radii, the area moment of inertia for bending around the z-axis is given according to [11, p. A7] by

$$I = h_\mathrm{s} t_\mathrm{s}^3 / 12. \tag{3.2}$$

In conclusion, the bending stiffness of a flat conductor is

$$S = E_0 I = E_0 \frac{h_\mathrm{s} t_\mathrm{s}^3}{12}. \tag{3.3}$$

Regarding CTCs, two assumptions are essential: 1. The transpositions between the strands are neglected. 2. The strands have a rectangular cross-section (the last part of this section gives an approach to take large edge radii into account).

Unbonded CTCs are some kind of laminated structure. The strands stacked on each other are not connected, and a transfer of shear between the strands does not occur. Therefore, sliding between strands takes place, and each strand bends independent (having its own neutral fiber). The CTC stiffness is equal to the sum of the strand stiffnesses, as derived in Equation (A.6). For a CTC with number N_s of strands, this results in

$$S = E_0 I N_\mathrm{s} = E_0 \frac{h_\mathrm{s} t_\mathrm{s}^3}{12} N_\mathrm{s}. \tag{3.4}$$

In the case of a bonded CTC, it is possible to assume a homogeneous cross-section. The underlying assumption is that the bonding withstands the shearing stresses arising between strands. It follows that sliding between strands does not take place. Two different cases arise, depending on the number of strands:

- If N_s is even, as given in the case of Figure 3.2, the cross-section with the outer dimensions $h_\mathrm{c}, t_\mathrm{c}$ is perfectly rectangular shaped. The neutral axis position is centered, and the standard formula for rectangular cross-sections according to Equation (3.2) is applicable. The resulting stiffness is given with the first case of Equation (3.5).
- If the bonded CTC has an odd number of strands, the neutral fiber position is no longer centered. [11, eq. (A.6)] provides an approach to calculate the axis position. The area moment of inertia results from the cross-sectional integral about this axis, as given in [11, eq. (A.7)]. The second case of Equation (3.5) shows the correlated result in the means of bending stiffness. For large N_s the term $6N_\mathrm{s} - 3/N_\mathrm{s}$ is negligible, leading to the same result as in the case of even strand number. Setting $N_\mathrm{s} = 1$ leads to the identical description as given for flat conductors with Equation (3.3).

$$S = \begin{cases} E_0 \dfrac{h_\mathrm{c} t_\mathrm{c}^3}{12} = E_0 \dfrac{2h_\mathrm{s}(\frac{N_\mathrm{s}}{2} t_\mathrm{s})^3}{12} = E_0 \dfrac{h_\mathrm{s} t_\mathrm{s}^3}{48} N_\mathrm{s}^3 & \text{for } N_\mathrm{s} \text{ even} \\ E_0 \dfrac{h_\mathrm{s} t_\mathrm{s}^3}{48} \left(N_\mathrm{s}^3 + 6N_\mathrm{s} - \dfrac{3}{N_\mathrm{s}} \right) & \text{for } N_\mathrm{s} \text{ odd} \end{cases} \tag{3.5}$$

Especially for CTCs with flat strands, the edge radii impact the area moment of inertia. The standard DIN EN 60317-0-2 [71] for the production of enameled rectangular copper wires allows a maximum possible edge radius of $t_s/2$ for thin conductors with $t_s \leq 1\,\mathrm{mm}$. With this as worst case assumption, the calculation of area moment of inertia is possible by splitting the strand into a rectangle and a circle, according to Figure 3.3. Using standard formulas for the area moment of inertia of a rectangle and a circle from [11], results in the total area moment of inertia according to

$$I = \frac{(h_s - t_s)t_s^3}{12} + \frac{\pi(t_s/2)^4}{4}. \tag{3.6}$$

The following approach is to describe the resulting stiffness reduction by an equivalent strand without edge radii, but with reduced height $\tilde{h}_s$:

$$I \stackrel{!}{=} \frac{\tilde{h}_s t_s^3}{12} \Rightarrow \tilde{h}_s = h_s - t_s\left(1 - \frac{3\pi}{16}\right) \approx h_s - 0.41 \cdot t_s \tag{3.7}$$

In conclusion, for CTCs made from thin strands with large edge radii, the strand height of an equivalent rectangular shaped strand reduces by 41 % of its thickness.

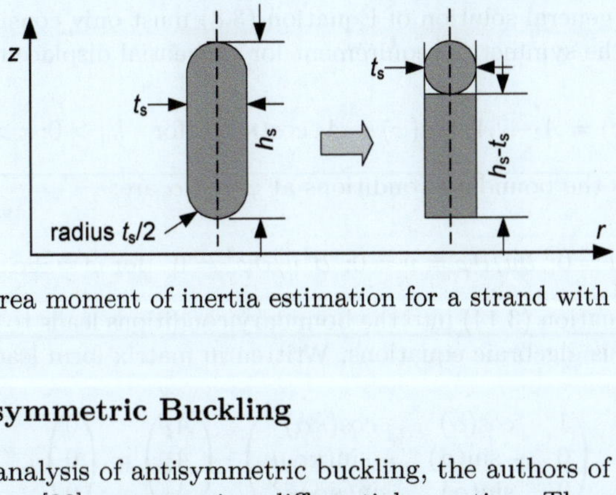

Figure 3.3: Area moment of inertia estimation for a strand with large edge radii

3.1.2 Antisymmetric Buckling

For the static analysis of antisymmetric buckling, the authors of [31, p. 207] use Lamb's equation [39] as governing differential equation. The governing Equation (3.8) is set up concerning the tangential displacement u. The underlying Euler-Bernoulli approach neglects shear deformation. The parameter k_q given in Equation (3.9) is defined in the means of bending radius R, Young's modulus E_0 and area moment of inertia I.

$$Ł[u] = \frac{\partial^6 u}{\partial \varphi^6} + 2\frac{\partial^4 u}{\partial \varphi^4} + \frac{\partial^2 u}{\partial \varphi^2} + k_q\left(\frac{\partial^4 u}{\partial \varphi^4} + \frac{\partial^2 u}{\partial \varphi^2}\right) = 0 \tag{3.8}$$

$$k_q = q_r R^3 / E_0 I \tag{3.9}$$

Equation (3.8) is valid for a circular arch with incompressible centerline, under a radial load q_r. The assumption of incompressibility leads to a direct correlation between radial displacement v and tangential displacement u as follows:

$$\epsilon = \frac{1}{R}\frac{\partial u}{\partial \varphi} - \frac{v}{R} = 0 \quad \rightarrow \quad \frac{\partial u}{\partial \varphi} = v \tag{3.10}$$

Taking the centerline compressibility into account leads to a much more cumbersome solution. In [9, p. 227] the authors present the solution for snap-through of compressible, shallow arches based on an energy approach.

According to [31, p. 208] the characteristic equation and correlated roots are:

$$\Lambda^6 + 2\Lambda^4 + \Lambda^2 + k_q(\Lambda^4 + \Lambda^2) = 0 \tag{3.11}$$

$$\Lambda_{1,2} = 0; \quad \Lambda_{3,4} = \pm i; \quad \Lambda_{5,6} = \pm i\kappa \tag{3.12}$$

$$\kappa = \sqrt{1 + k_q} \tag{3.13}$$

Generally, an antisymmetric solution with respect to radial displacement $v(\varphi)$ corresponds to a symmetric solution based on tangential displacement $u(\varphi)$. This suggests that a general solution of Equation (3.8) must only consist of even functions, to fulfill the symmetric requirement for tangential displacement:

$$u(\varphi) = A_1 + A_2\cos(\varphi) + A_3\cos(\kappa\varphi) \quad \text{for} \quad k_q > 0; \kappa > 1 \tag{3.14}$$

For hinged ends the boundary conditions at $\varphi = \pm\alpha$ are:

$$u = 0; \; \partial u/\partial\varphi = v = 0; \; \partial^3 u/\partial\varphi^3 = 0 \quad \text{for } \varphi = \pm\alpha. \tag{3.15}$$

Substituting Equation (3.14) into the boundary conditions leads to a linear system of 3 homogeneous algebraic equations. Written in matrix form leads to

$$\begin{pmatrix} 1 & \cos(\alpha) & \cos(\kappa\alpha) \\ 0 & -\sin(\alpha) & -\sin(\kappa\alpha)\kappa \\ 0 & \sin(\alpha) & \sin(\kappa\alpha)\kappa^3 \end{pmatrix} \cdot \begin{pmatrix} A_1 \\ A_2 \\ A_3 \end{pmatrix} = \begin{pmatrix} 0 \\ 0 \\ 0 \end{pmatrix}. \tag{3.16}$$

For such a cubic homogeneous equation system the determinant of the coefficient matrix determines the solubility: If it is different from zero the system has only one solution, which is for this kind of homogeneous linear system the trivial solution $A_1 = A_2 = A_3 = 0$. If the determinant is zero, and also all three auxiliary determinants are zero, the system has an infinite number of solutions. For the auxiliary determinants, this requirement is fulfilled for all values of α and κ. Setting the coefficient matrix determinant to zero results in

$$-\sin(\kappa\alpha)\sin(\alpha)\kappa(\kappa^2 - 1) = 0. \tag{3.17}$$

As $\kappa > 1$, only the factor $\sin(\kappa\alpha)$ can force Equation (3.17) to get zero. This leads to

$$\sin(\kappa\alpha) = 0 \quad \Rightarrow \quad \kappa = n\pi/\alpha = nN_{\mathrm{as}}/2 \quad \text{for } n \geq 1 \tag{3.18}$$

with $N_{\mathrm{as}} = 2\pi/\alpha$ being the number of axial sticks supporting the winding in the radial direction.

Substituting Equation (3.18) with $n = 1$ into Equation (3.13) and replacing k_{q} inside Equation (3.9) results in the lowest critical buckling load:

$$q_{\mathrm{cr}} = \frac{E_0 I}{R^3} k_{\mathrm{q}} = \frac{E_0 I}{R^3}(\kappa^2 - 1) = \frac{E_0 I}{R^3}\left(\left(\frac{N_{\mathrm{as}}}{2}\right)^2 - 1\right). \tag{3.19}$$

Derived from the pipe formula [8, p. 215], the critical compressive stress σ_{cr}, which appears normal to the conductor cross-section, depends on q_{cr} by the relation

$$\sigma_{\mathrm{cr}} = \frac{R}{A} q_{\mathrm{cr}}. \tag{3.20}$$

Substituting Equation (3.19) into Equation (3.20) results in

$$\sigma_{\mathrm{cr}} = \frac{E_0 I}{4AR^2}\left(N_{\mathrm{as}}^2 - 4\right). \tag{3.21}$$

In the case of a solid conductor, Equation (3.3) defines the area moment of inertia I, and the cross-sectional area is given by $A = t_{\mathrm{s}} h_{\mathrm{s}}$. Introducing the mean winding diameter $D_{\mathrm{m}} = 2R$ and the conductor slenderness ratio according to Equation (3.23) gives

$$\sigma_{\mathrm{cr}} = \frac{E_0}{12}\left(\frac{t_{\mathrm{s}} N_{\mathrm{as}}}{D_{\mathrm{m}}}\right)^2 \left(\frac{N_{\mathrm{as}}^2 - 4}{N_{\mathrm{as}}^2}\right) = \frac{E_0 T^2}{12}\left(1 - \frac{4}{N_{\mathrm{as}}^2}\right). \tag{3.22}$$

$$T = t_{\mathrm{s}} N_{\mathrm{as}}/D_{\mathrm{m}} \tag{3.23}$$

For an unbonded CTC with $I = N_{\mathrm{s}} h_{\mathrm{s}} t_{\mathrm{s}}^3/12$ and $A = N_{\mathrm{s}} h_{\mathrm{s}} t_{\mathrm{s}}$ Equation (3.21) results in the same solution, as the number of strands N_{s} vanishes. Only the critical load q_{cr} according to Equation (3.20) is N_{s}-times higher for a CTC, compared to a single conductor with the same outer strand dimensions.

3.1.3 Symmetric Buckling

In the case of symmetric buckling it is sufficient to regard one sector with the angular dimension of α. Fixed supports restrain the winding endings. The buckling shape with the lowest critical load correlates once again to an antisymmetric mode, but this time based on the half angular extension compared to Section 3.1.2. In conclusion, the governing equation on tangential displacement and the solution approach $u(\varphi)$ presented in the previous section are applicable.

For fixed ends the boundary conditions at $\varphi = \pm\alpha/2$ are

$$u = 0,\ \partial u/\partial\varphi = v = 0,\ \partial u^2/\partial\varphi^2 = \partial v/\partial\varphi = 0. \tag{3.24}$$

Proceeding as before, subjecting Equation (3.14) into Equation (3.24) leads to the following equation system given in matrix form:

$$\begin{pmatrix} 1 & \cos(\alpha/2) & \cos(\kappa\alpha/2) \\ 0 & -\sin(\alpha/2) & -\sin(\kappa\alpha/2)\kappa \\ 0 & -\cos(\alpha/2) & -\cos(\kappa\alpha/2)\kappa^2 \end{pmatrix} \cdot \begin{pmatrix} A_1 \\ A_2 \\ A_3 \end{pmatrix} = \begin{pmatrix} 0 \\ 0 \\ 0 \end{pmatrix} \tag{3.25}$$

The final equation for the critical parameter κ is

$$\tan(\alpha/2)\kappa = \tan(\alpha/2\kappa). \tag{3.26}$$

A closed-form representation for κ is not possible, which is why Table 3.1 provides the solutions for some exemplary sector angles α. Equation (3.19) leads to the critical load. Based on Equation (3.20) the critical stress results in

$$\sigma_{\mathrm{cr}} = \frac{E_0 I}{AR^2}\left(\kappa^2 - 1\right). \tag{3.27}$$

The substitution of I and A in the means of cross-sectional parameters, and the implementation of slenderness ratio T according to Equation (3.23) and mean diameter $D_{\mathrm{m}} = 2R$ lead to a critical stress of

$$\sigma_{\mathrm{cr}} = \frac{E_0 T^2}{12}\left(\frac{4\kappa^2 - 4}{N_{\mathrm{as}}^2}\right). \tag{3.28}$$

Table 3.1: Parameter κ for the critical load of asymmetric and symmetric buckling

α in °	45	30	24	20	15	12	8	4
N_{as}	8	12	15	18	24	30	40	90
κ (asymm.)	4	6	7.5	9	12	15	20	45
κ (symm.)	11.47	17.18	21.47	25.75	34.33	42.92	64.37	128.73

Comparing the values of κ listed in Table 3.1 reveals a factor of 2.86 between antisymmetric and symmetric case. Regarding Equation (3.19) the critical load is approximately $2.86^2 = 8.18$ times higher in the symmetric case. However, this is only possible for an ideal rigid fixation at the spacer positions.

3.1.4 Transition from Buckling to Pure Bending

The authors of [58] investigated the character of the buckling behavior for shallow circular arches under radial pressure. One of their major conclusions is: The occurrence of instability buckling depends on the ratio between rise and half thickness of the arch. This ratio has been declared as dimensionless rise $h_r^* = h_r/(2t_s)$, as derived in Equation (A.18). The authors found out, that for very shallow arches no instability buckling occurs. In this case, the arch acts as a nonlinear spring, showing a smooth transition from the original to the buckled state.

- For fixed ends no buckling occurs for $h_r^* < 2.85$.
- For hinged ends the authors of [19, p. 153] estimated that if $h_r^* < 1.11$, no buckling occurs.

The case of pure bending still is problematic with respect to transformers, as this phenomenon also will result in residual winding deformation. Examples of such behavior are given in Section 4.5.4 for thermosetting CTCs.

3.1.5 Involving Elastoplastic Properties

Taking elastoplastic copper properties into account is possible by replacing the linear Young's modulus E_0 by a nonlinear tangent modulus $E_t(\sigma)$, as defined in the Ramberg Osgood representation according to Equation (2.34).
For the case of antisymmetric buckling and with the assumption of $\sigma_{cr} \leq \sigma_{0.2}$, the substitution of $E_0 = E_t(\sigma)$ in Equation (3.22) results in the implicit equation:

$$\sigma_{cr} = \frac{E_t(\sigma_{cr})T^2}{12}\left(1 - \frac{4}{N_{as}^2}\right) = \frac{E_0 T^2}{12\left(1 + \frac{3}{7}m\left(\frac{\sigma_{cr}}{\sigma_0}\right)^{m-1}\right)}\left(1 - \frac{4}{N_{as}^2}\right) \quad (3.29)$$

For the resulting implicit equation, only a numerical solution is possible. A closed-form representation of σ_{cr} is not feasible. To get rid of the underlying case distinction whether the stress is below or above $\sigma_{0.2}$, the following approach is helpful: An initial calculation uses the tangent modulus definition for $\sigma \leq \sigma_{0.2}$. If the maximum stress stays below the 0.2 % proof stress, the calculation is finished. Otherwise a second calculation based on the formulation of Equation (2.34) for $\sigma > \sigma_{0.2}$ is necessary.
Finally, the critical load respecting material nonlinearity results from the substitution of σ_{cr} into Equation (3.20). For the case of unbonded CTCs, A refers to the whole cross-sectional area $A = N_s h_s t_s$.

Section 3.2.3.3 compares some results from Equation (3.29) to FEA results.

3.1.6 Dynamic Buckling Analysis

3.1.6.1 Hydrodynamic Damping and Inertial Mass

Inside oil-immersed transformers, the oil may affect the amplitude of conductor vibrations during normal load case, as well as failure modes like buckling. This impact strongly depends on coil structure, spacers and oil channels. A detailed analysis of hydrodynamic effects is only possible with complex fluid-structure coupled simulations. However, this section proposes a simplified analytic approach for the dynamic radial buckling of rectangular shaped conductors, supported by axial sticks. The following assumptions for conductor and oil as fluid are inevitable:

- The oil is incompressible with density ρ_f and dynamic viscosity μ_f.
- There are no surrounding effects like barriers. The fluid is unbounded.
- The length of the curved conductor exceeds its cross-sectional dimensions.

Generally, a fluid damping force and a fluid inertial force contribute to hydrodynamic damping. The fluid damping force results from the viscous drag of a fluid. It is related linearly to the velocity of the moving body, and is out-of-phase with its oscillation. The fluid inertial force is correlated to the fluid mass which gets displaced by the moving object. It is related linearly to the acceleration of the moving body, and is in-phase with its oscillation. In the following the damping part is neglected as it is very difficult to establish a fixed value, especially for multiple member structures, and experimental tests are necessary [18].

The fluid inertial force is characterized by the hydrodynamic mass coefficient per unit length k_m ($\mathrm{kg\,m^{-1}}$). It depends on the cross-section of the moving object, its moving speed and the density and viscosity of the fluid. An analytical calculation is possible for small deflections with respect to the cross-sectional dimensions. This case neglects all nonlinear, convective inertial effects [2].

The authors of [15] calculated hydrodynamic mass coefficients for rectangular shaped harmonic vibrating microcantilevers. They combined an exact analytical solution based on Stokes equations (which is valid for a flat cantilever), with correction factors deduced from numerical analysis. Edge and thickness effects have been taken into account. This approach is valid for a wide range of aspect ratios between 1/56 ... 1 for rectangular shaped cross-sections. The hydrodynamic hydrodynamic mass coefficient is defined by

$$k_\mathrm{m} = \frac{\rho_\mathrm{f} t_\mathrm{c}^2 C_\mathrm{m}}{\sqrt{2Re}} \tag{3.30}$$

with fluid density ρ_f, conductor width t_c in radial movement direction, Reynolds number Re and the numerical correction factor C_m. The Reynolds number indicates the relationship between viscous and inertial forces in the fluid. It is given to

$$Re = \frac{\omega t_\mathrm{c}^2 \rho_\mathrm{f}}{4\mu_\mathrm{f}} \tag{3.31}$$

with dynamic viscosity μ_f and angular frequency ω of the exciting load. In the case of transformer windings subjected to an electromagnetic load according to Equation (2.17), ω equals the double power grid angular frequency $2\omega_0$. The correction factor depends on cross-section aspect ratio h_c/t_c and the Reynolds number, as given in Equation (3.32). This approach is valid for $1/56 \leq h_c/t_c \leq 1$ and $10 \leq Re \leq 10^4$ [15].

$$C_m = 1.658 \left(\frac{h_c}{t_c}\right)^{1.83} \sqrt{Re} + 3.08 \left(\frac{h_c}{t_c}\right)^{0.85} + 1 \tag{3.32}$$

A second approach is given in [18]. The author derives the added mass coefficient k_m for different aspect ratios of rectangular shaped bodies. It only depends on the height h_c perpendicular to the movement direction and fluid density ρ_f:

$$k_m = 1.98\pi\rho_f h_c^2 \quad \text{for} \quad h_c/t_c = 1/5 \tag{3.33a}$$

$$k_m = 1.51\pi\rho_f h_c^2 \quad \text{for} \quad h_c/t_c = 1 \tag{3.33b}$$

$$k_m = 1.14\pi\rho_f h_c^2 \quad \text{for} \quad h_c/t_c = 10 \tag{3.33c}$$

The following comparison assumes three different conductors vibrating in transformer oil. The dimensions are given in Figure 3.4. Both dynamic viscosity $\mu_f = 6.94 \cdot 10^{-3}\,\mathrm{kg\,m^{-1}\,s^{-1}}$ and density $\rho_f = 866.7\,\mathrm{kg\,m^{-3}}$ are derived from the datasheet values for Shell Diala D transformer oil at a temperature of 40 °C. The harmonic conductor vibration has a angular frequency of $\omega = 2\omega_0 = 2\pi \cdot 100\,\mathrm{Hz}$. Furthermore, the conductor mass per unit length is

$$m_c = \rho_{co} h_c t_c \tag{3.34}$$

with the copper density $\rho_{co} = 8930\,\mathrm{kg\,m^{-3}}$.

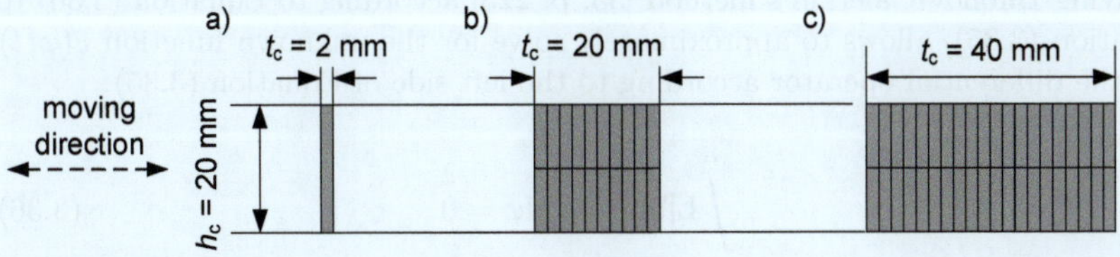

Figure 3.4: Cross-sections for conductor type a), b) and c) for damped dynamic buckling analysis

Table 3.2 lists the resulting hydrodynamic mass coefficients for the three conductors. The hydrodynamic mass per unit length coefficients k_m show a reasonable coincidence between the approaches of [15] and [18]. However, the approach according to [15] is more precise, as the included numerical correction factors respect edge and thickness effects. The contribution of hydrodynamic mass only is significant for the thin conductor type a), where $m_c \approx k_m$. For the remaining two conductors the hydrodynamic mass is less than 10 % of the conductor mass.

Table 3.2: Conductor mass and hydrodynamic mass coefficients

Conductor type	m_c in $\frac{kg}{m}$	k_m in $\frac{kg}{m}$ Equation (3.30) [15]	k_m in $\frac{kg}{m}$ Equation (3.33) [18]
a)	0.357	0.281	0.310
b)	3.572	0.417	0.411
c)	7.144	0.472	0.463

3.1.6.2 Governing Equation

The governing Equation (3.35) of vibration for a two-hinged arch is taken from [31, p. 375]. It is based on Lamb's equation [39], which has been already used for the static approach in Section 3.1.2. Additional parameters are the conductor mass per unit length m_c and the time dependent radial line load $q_r(t)$, as defined in Equation (2.17). The hydrodynamic mass coefficient k_m contributes to m_c.

$$\begin{aligned} Ł[u] =& \frac{\partial^6 u}{\partial \varphi^6} + 2\frac{\partial^4 u}{\partial \varphi^4} + \frac{\partial^2 u}{\partial \varphi^2} + \frac{q_r(t) R^3}{E_0 I}\left(\frac{\partial^4 u}{\partial \varphi^4} + \frac{\partial^2 u}{\partial \varphi^2}\right) \\ &+ \frac{(m_c + k_m) R^4}{E_0 I}\frac{\partial^2}{\partial t^2}\left(\frac{\partial^2 u}{\partial \varphi^2} - u\right) = 0 \end{aligned} \tag{3.35}$$

Applying Bubnov-Galerkin's method [35, p. 223] according to Equation (3.36) to Equation (3.35), allows to approximately solve for the unknown function $u(\varphi, t)$. Ł is the differential operator according to the left side of Equation (3.35).

$$\int_{-\alpha}^{\alpha} Ł[u] u(t, \varphi) \mathrm{d}\varphi = 0 \tag{3.36}$$

Equation (3.37) describes an approximate solution for the tangential displacement $u(\varphi, t)$. This approach satisfies the boundary conditions of a two-hinged arch given by Equation (3.15). The function $f(t)$ only depends on time, and has to be determined by Bubnov-Galerkin's method.

$$u(\varphi, t) = -f(t)\frac{\alpha}{\pi}\left(1 + \cos\left(\frac{\pi\varphi}{\alpha}\right)\right) \tag{3.37}$$

The radial deformation $v(\varphi,t) = \partial u(\varphi,t)/\partial\varphi = f(t)\sin(\frac{\pi\varphi}{\alpha})$ verifies the antisymmetric, sinusoidal shape of vibration.
Integration of Equation (3.36) with replacement of $\xi = \pi/\alpha = N_{\mathrm{as}}/2$ leads to Equation (3.38). The detailed steps of solving the Galerkin integral are given in Appendix A.2.

$$\begin{aligned} &\frac{(m_{\mathrm{c}}+k_{\mathrm{m}})R^4}{E_0 I}\left(\xi^2+3\right)f''(t) \\ &+\left(\frac{q_{\mathrm{r}}(t)R^3}{E_0 I}\left(\xi^2-\xi^4\right)+\xi^6-2\xi^4+\xi^2\right)f(t)=0. \end{aligned} \tag{3.38}$$

Replacement of

$$\omega_{\mathrm{cr}} = \frac{\xi(\xi^2-1)}{\sqrt{\xi^2+3}}\frac{1}{R^2}\sqrt{\frac{E_0 I}{m_{\mathrm{c}}+k_{\mathrm{m}}}} \tag{3.39}$$

$$\gamma = \frac{\xi^4-\xi^2}{\xi^2+3} \tag{3.40}$$

transforms Equation (3.38) to

$$f''(t)+\omega_{\mathrm{cr}}^2\left(1-\frac{q_{\mathrm{r}}(t)}{(m_{\mathrm{c}}+k_{\mathrm{m}})R\omega_{\mathrm{cr}}^2}\gamma\right)f(t)=0. \tag{3.41}$$

Equation (3.41) contains the static critical load q_{cr}, which is defined by

$$q_{\mathrm{cr}} = \frac{(m_{\mathrm{c}}+k_{\mathrm{m}})R\omega_{\mathrm{cr}}^2}{\gamma} = \frac{\xi^2\left(\xi^2-1\right)^2}{\xi^4-\xi^2}\frac{E_0 I}{R^3} = \frac{E_0 I}{R^3}\left(\xi^2-1\right) \tag{3.42}$$

This representation is equal to the solution of static buckling analysis according to Equation (3.19), if $\xi = N_{\mathrm{as}}/2$ is substituted. The final differential equation is given by

$$f''(t)+\omega_{\mathrm{cr}}^2\left(1-\frac{q_{\mathrm{r}}(t)}{q_{\mathrm{cr}}}\right)f(t)=0. \tag{3.43}$$

It has the form of a Hill equation [64]. A general solution for $f(t)$ in dependency of geometric parameters and $q_{\mathrm{r}}(t)$ is not possible. Therefore, the next section proposes a stability analysis.

3.1.6.3 Mathieu Equation

The harmonic excitation given by $q_r(t)$ allows converting the Hill equation into a Mathieu equation [64]. Insertion of $q_r(t)$ according to Equation (2.17) into Equation (3.43) leads to

$$f''(t) + \omega_{cr}^2 \left(1 - \frac{\hat{q}_r}{2q_{cr}} + \frac{\hat{q}_r}{2q_{cr}} \cos(2\omega_0 t)\right) f(t) = 0. \tag{3.44}$$

Substituting $\tau = \omega_0 t$ results in

$$f''(\tau) + \frac{\omega_{cr}^2}{\omega_0^2} \left(1 - \frac{\hat{q}_r}{2q_{cr}} + \frac{\hat{q}_r}{2q_{cr}} \cos(2\tau)\right) f(\tau) = 0. \tag{3.45}$$

The use of the normalized parameters ν and ψ according to

$$\nu = \left(\frac{\omega_{cr}}{\omega_0}\right)^2 \left(1 - \frac{\hat{q}_r}{2q_{cr}}\right) \tag{3.46}$$

$$\psi = \left(\frac{\omega_{cr}}{\omega_0}\right)^2 \frac{\hat{q}_r}{2q_{cr}} \tag{3.47}$$

transforms Equation (3.45) into the standard form of a Mathieu equation:

$$f''(\tau) + (\nu + \psi \cos(2\tau))\, f(\tau) = 0 \tag{3.48}$$

3.1.6.4 Stability Analysis

The parameters ν, ψ define whether the solution of Equation (3.48) is limited or tends to infinity. The latter corresponds to an unstable behavior. The Ince-Strutt stability chart [31, p. 373] is a suitable diagram to visualize all stability boundaries in the $\nu\psi$-parameter-plane.

For the present Mathieu equation these boundaries $f(\tau)$ correspond to periodic solutions [64, p. 56]. The following Fourier series describes these solutions:

$$f(\tau) = a_0 + \sum_{n=1}^{\infty} \left[a_n \cos(n\tau) + b_n \sin(n\tau)\right] \tag{3.49}$$

Applying the principle of harmonic balance determines the parameters a_0, a_n, b_n. The substitution of Equation (3.49) into Equation (3.48) results in an equation system of infinite dimension. As only odd and even cosine and sine terms are involved it is possible to group the resulting equations in four independent sets.

Odd cosines:

$$\begin{array}{l} \cos(1\tau): \\ \cos(3\tau): \\ \cos(5\tau): \\ \cos(7\tau): \\ \\ \end{array} \begin{pmatrix} \frac{\psi}{2}+\nu-1 & \frac{\psi}{2} & 0 & 0 & \dots \\ \frac{\psi}{2} & \nu-3^2 & \frac{\psi}{2} & 0 & \dots \\ 0 & \frac{\psi}{2} & \nu-5^2 & \frac{\psi}{2} & \dots \\ 0 & 0 & \frac{\psi}{2} & \nu-7^2 & \dots \\ \vdots & \vdots & \vdots & \vdots & \ddots \end{pmatrix} \cdot \begin{pmatrix} a_1 \\ a_3 \\ a_5 \\ a_7 \\ \vdots \end{pmatrix} = \begin{pmatrix} 0 \\ 0 \\ 0 \\ 0 \\ \vdots \end{pmatrix} \quad (3.50)$$

Even cosines:

$$\begin{array}{l} \cos(0\tau): \\ \cos(2\tau): \\ \cos(4\tau): \\ \cos(6\tau) \\ \\ \end{array} \begin{pmatrix} \nu & \frac{\psi}{2} & 0 & 0 & \dots \\ \frac{\psi}{2} & \nu-2^2 & \frac{\psi}{2} & 0 & \\ 0 & \frac{\psi}{2} & \nu-4^2 & \frac{\psi}{2} & \dots \\ 0 & 0 & \frac{\psi}{2} & \nu-6^2 & \\ \vdots & \vdots & \vdots & \vdots & \ddots \end{pmatrix} \begin{pmatrix} a_0 \\ a_2 \\ a_4 \\ a_6 \\ \vdots \end{pmatrix} = \begin{pmatrix} 0 \\ 0 \\ 0 \\ 0 \\ \vdots \end{pmatrix} \quad (3.51)$$

Even sines:

$$\begin{array}{l} \sin(1\tau): \\ \sin(3\tau): \\ \sin(5\tau) \\ \sin(7\tau): \\ \\ \end{array} \begin{pmatrix} -\frac{\psi}{2}+\nu-1 & \frac{\psi}{2} & 0 & 0 & \dots \\ \frac{\psi}{2} & \nu-3^2 & \frac{\psi}{2} & 0 & \dots \\ 0 & \frac{\psi}{2} & \nu-5^2 & \frac{\psi}{2} & \dots \\ 0 & 0 & \frac{\psi}{2} & \nu-7^2 & \dots \\ \vdots & \vdots & \vdots & \vdots & \ddots \end{pmatrix} \begin{pmatrix} b_1 \\ b_3 \\ b_5 \\ b_7 \\ \vdots \end{pmatrix} = \begin{pmatrix} 0 \\ 0 \\ 0 \\ 0 \\ \vdots \end{pmatrix} \quad (3.52)$$

Odd sines:

$$\begin{matrix} \sin(2\tau): \\ \sin(4\tau): \\ \sin(6\tau): \\ \sin(8\tau): \\ \\ \end{matrix} \begin{pmatrix} \nu - 2^2 & \frac{\psi}{2} & 0 & 0 & \dots \\ \frac{\psi}{2} & \nu - 4^2 & \frac{\psi}{2} & 0 & \dots \\ 0 & \frac{\psi}{2} & \nu - 6^2 & \frac{\psi}{2} & \dots \\ 0 & 0 & \frac{\psi}{2} & \nu - 8^2 & \dots \\ \vdots & \vdots & \vdots & \vdots & \ddots \end{pmatrix} \cdot \begin{pmatrix} b_2 \\ b_4 \\ b_6 \\ b_8 \\ \vdots \end{pmatrix} = \begin{pmatrix} 0 \\ 0 \\ 0 \\ 0 \\ \vdots \end{pmatrix} \tag{3.53}$$

The four underlying homogeneous equation systems have a non-trivial solution if the determinants are zero. This formulation corresponds to an eigenvalue problem, as ν only appears in the main diagonal of each matrix. The eigenvalues only depend on ψ and a visualization in the $\nu\psi$-parameter plane is possible. The resulting diagram corresponds to an Ince-Strutt stability diagram with boundary curves separating the area into stable and unstable regions. A limitation to a finite number of equations allows calculating a numeric solution.

Figure 3.5 visualizes an exemplary evaluation with a dimension of 30 equations. White areas correspond to unstable regions: The related parameter sets (ν, ψ) lead to instability or in the case of transformer windings to buckling. The gray regions are stable. The corresponding (ν, ψ) cause bounded deflection and no buckling.

Three exemplary parameter sets of $\hat{q}_\mathrm{r}$ and ω_0 are drawn by the means of red lines. The first corresponds to a load amplitude of $\hat{q}_\mathrm{r} = q_\mathrm{cr}$, the second to $\hat{q}_\mathrm{r} = 0.5 \cdot q_\mathrm{cr}$ and the third to $\hat{q}_\mathrm{r} = 0.05 \cdot q_\mathrm{cr}$. The length of both lines is correlated to an excitation frequency variation between $10 \dots 100\,\%$ of ω_cr.
It is particularly noticeable that the line corresponding to the critical load does not exclusively run through unstable areas. Especially for high frequencies close to ω_cr it ends in a stable region. For frequencies smaller than $0.8 \cdot \omega_\mathrm{cr}$ the parameter line runs close to the boundary between stable and instable regions.
For reduced load of $0.5 \cdot q_\mathrm{cr}$ the underlying excitation frequencies are stable to the largest extent. The only unstable region is passed between $0.7 \dots 0.9 \cdot \omega_\mathrm{cr}$. With decreasing load the instable frequency range narrows to $\nu = 1$, which leads according to Equation (3.48) to $\omega_0 = \omega_\mathrm{cr}$.
This indicates that the critical frequency defined by Equation (3.39) is not necessarily an unstable resonance frequency. The load, which is not purely harmonic, but has a constant offset of $\hat{q}_\mathrm{r}/2$, has impact on the unstable frequencies.
Section 3.2.3.4 explains the application of the Ince-Strutt chart by the means of a simulation model in more detail. The black cross in Figure 3.5 marks the underlying stability parameters.

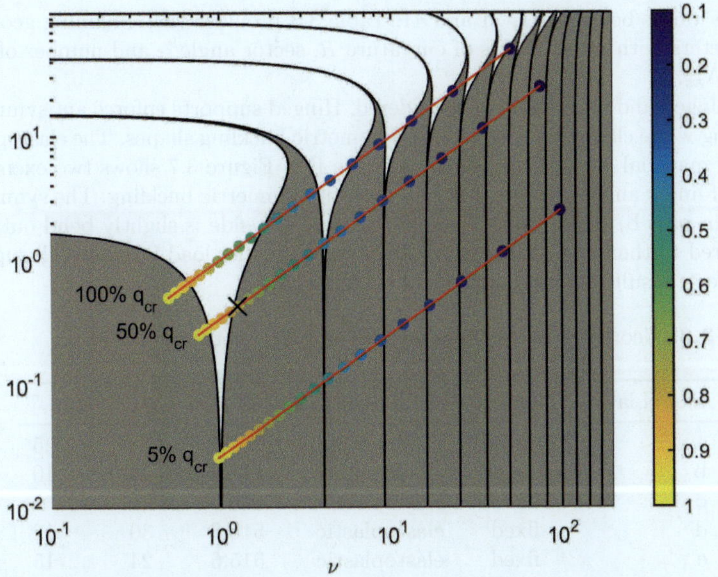

Figure 3.5: Ince-Strutt stability chart for Mathieu Equation (3.48) with exemplary parameter sets for $\hat{q}_r$ and ω_0 (red lines).

3.2 Finite Element-Based Analysis

3.2.1 Simplified CTC Model

A detailed representation of transposed conductors in FEA studies is very challenging: Especially the transposing strands complicate 3D geometry creation and lead to complex contact areas between the thin and vast strand surfaces. The main purpose of a simplified CTC model is to reduce model complexity. Figure 3.6 visualizes both a detailed and simplified CTC model for the example of CTC-B-21. The simplified model has no transpositions, and the strands remain on their radial position. Merging two adjacent strands in both stacks to one strand with the double height reduces the contact areas further. This kind of simplification is only suitable for simulating the bending behavior in the radial direction.

The author's publications [80, 83] describe a detailed comparison between complex and simplified model which are the basis for the subsequent investigations. The fundamental method is to compare load-deflection characteristics and critical loads deduced from static structural FEA based bending studies. Furthermore, the simulation results verify the accuracy of analytic results based on Section 3.1. The geometry refers to curved CTC sectors which are loaded by a volume correlated radial force. The strand dimensions are from CTC-B-21. The number N_s of

strands varies between 11, 21 and 41. Table 3.3 provides the remaining geometry parameters with mean radius of curvature R, sector angle α and number of axial sticks N_{as}.
Both hinged and fixed ends are considered. Hinged supports enforce antisymmetric buckling while clamped ends result in symmetric buckling shapes. The elastoplastic copper material is co-65 according to Table B.1. Figure 3.7 shows two exemplary cases of linear antisymmetric and elastoplastic symmetric buckling. The symmetric shape in part b) is not exact symmetric as the left side is slightly bend outwards compared to the main deformation direction. A further load increase will suppress this effect, resulting in a symmetric buckling shape.

Table 3.3: Geometry parameters for curved CTC sectors

model label	support	FEA study	R in mm	N_{as}	α in °
a	hinged	linear	687.5	48	7.5
b	hinged	linear	515.6	36	10
c	fixed	elastoplastic	515.6	40	9
d	fixed	elastoplastic	515.6	30	12
e	fixed	elastoplastic	515.6	24	15
f	fixed	elastoplastic	515.6	18	20

Figure 3.8 a) visualizes the relative error between detailed and simplified FEA. Part b) provides the relative error between detailed FEA and analytic approach. The underlying equations for relative error are

$$\text{error} = \frac{\text{simplified} - \text{detailed}}{\text{detailed}} \cdot 100\,\% \quad \left(= \frac{\text{analytic} - \text{detailed}}{\text{detailed}} \cdot 100\,\% \right). \qquad (3.54)$$

Negative values indicate that the results gathered from the simplified FEA models respective the analytic equations are lower, which is some kind of safe-side estimation. The real buckling strength can be expected to be higher than the values derived from the simplified approaches.
In the case of linear, antisymmetric buckling (model types a, b) the biggest deviation occur for the 11 stranded models with around $-24\,\%$. An increasing number of strands reduces this deviation especially for the simplified FEA model. The relative error deduced from the analytic approach stays between $-10\ldots-20\,\%$. To conclude, in the case of linear buckling the transpositions slightly contribute to the conductor stiffness by an increased buckling load compared to the simplified approaches. The error in critical load for elastoplastic symmetric buckling is much smaller for the models c, d and e. Only for model f with the smallest slenderness ratio, the deviation especially for the analytic approach increases. Also in this case, the simplified models are a safe-side estimation as the deviation to the detailed model including transpositions are almost always negative.

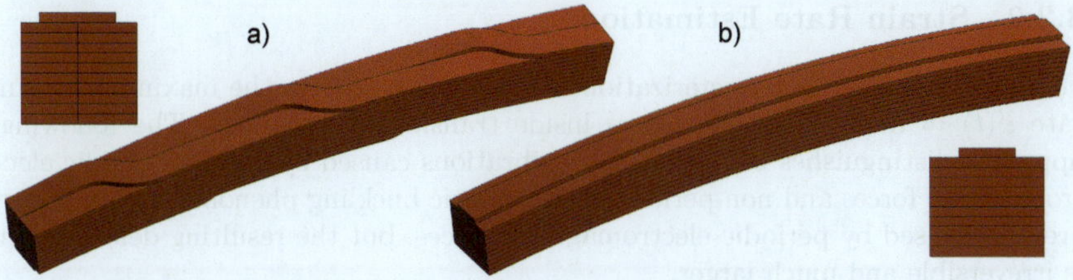

Figure 3.6: a) Detailed CTC model for radial buckling analysis
b) Simplified CTC model for radial buckling analysis

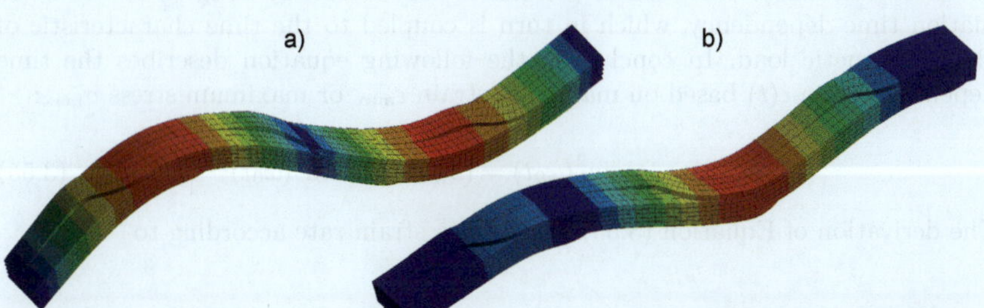

Figure 3.7: a) Detailed model b (11 strands, linear antisymmetric buckling)
b) Detailed model f (11 strands, elastoplastic symmetric buckling)

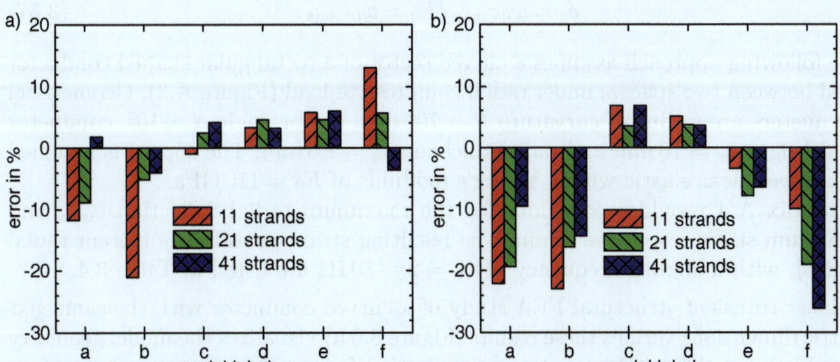

Figure 3.8: Relative deviation in critical buckling loads
a) Detailed vs. simplified FEA model
b) Detailed FEA vs. analytic approach

3.2.2 Strain Rate Estimation

For reliable material characterization it is evident to assess the maximum strain rate $\epsilon'(t) = \mathrm{d}\epsilon(t)/\mathrm{d}t$ which occurs inside transformer windings. The following approach distinguishes between elastic vibrations caused by 100 Hz periodic electromagnetic forces and non-periodic, elastoplastic buckling phenomena. The latter are also caused by periodic electromagnetic forces, but the resulting deformation is irreversible and much larger.

3.2.2.1 Elastic Vibration

In the linear elastic case time dependent strain is directly coupled to the deformation time-dependency, which in turn is coupled to the time characteristic of electromagnetic load. In conclusion, the following equation describes the time depended strain $\epsilon(t)$ based on maximum strain $\epsilon_{\max}$ or maximum stress $\sigma_{\max}$:

$$\epsilon(t) = \epsilon_{\max} \sin^2(\omega_0 t) = (\sigma_{\max}/E_0) \sin^2(\omega_0 t) \tag{3.55}$$

The derivation of Equation (3.55) results in a strain rate according to

$$\begin{aligned} \epsilon'(t) &= (2\omega_0\sigma_{\max}/E_0) \sin(\omega_0 t) \cos(\omega_0 t) \\ &= (\omega_0\sigma_{\max}/E_0) \sin(2\omega_0 t) = \delta_{\mathrm{s}} \sin(2\omega_0 t) \end{aligned} \tag{3.56}$$

with the strain rate amplitude

$$\delta_{\mathrm{s}} = \omega_0\sigma_{\max}/E_0 = \omega_0\epsilon_{\max}. \tag{3.57}$$

The following approach assumes a curved sector of a rectangular shaped conductor fixed between two spacers under radial compressive load (Figure A.3). Geometrical parameters are radius of curvature $R = 700\,\mathrm{mm}$, sector angle $\alpha = 10°$, conductor axial height $h_{\mathrm{s}} = 10\,\mathrm{mm}$ and radial thickness $t_{\mathrm{s}} = 1.5\,\mathrm{mm}$. The copper is assumed to behave linear elastic with a Young's modulus of $E_0 = 110\,\mathrm{GPa}$.
Appendix A.4 provides equations for the maximum radial deflection $v_{\max}$ and maximum stress respective strain. The resulting strain rates for 3 different radial loads $\hat{q}_{\mathrm{r}}$ with a angular frequency of $\omega_0 = 2\pi \cdot 50\,\mathrm{Hz}$ are listed in Table 3.4.

A linear transient structural FEA study of a curved conductor with the same geometric dimensions verifies these results. Figure 3.9 a) visualizes the model geometry. Part b) shows the maximum strain distribution for the time step correlating to a load of $600\,\mathrm{N\,m^{-1}}$. Figure 3.10 visualizes the time behavior of maximum deflection, maximum strain and strain rate for the load amplitudes according to Table 3.4: Both maximum deformation and strain follow the shape of exciting load (dashed line). The strain rate shown in Figure 3.10 c) has a slightly distorted sinusoidal shape with double angular frequency $2\omega_0$ as derived in Equation (3.56).

The comparison between analytic and FEA results leads to a good agreement for maximum strain and strain rate. Only $v_{\max}$ slightly differs from the FEA results. Generally, in the case of elastic vibrations, $\sigma_{\max}$ has to stay beyond the yield strength. Therefore, it is possible to estimate δ_s by Equation (3.57). Assuming a worst case scenario with $E_0 = 110\,\text{GPa}$, $\sigma_{\max} = 200\,\text{MPa}$ and $\omega_0 = 2\pi \cdot 50\,\text{Hz}$ leads to $\delta_s = 0.57\,\text{s}^{-1}$. Regarding Figure 2.11 a quasi static copper characterization is sufficient for the elastic analysis of transformer windings.

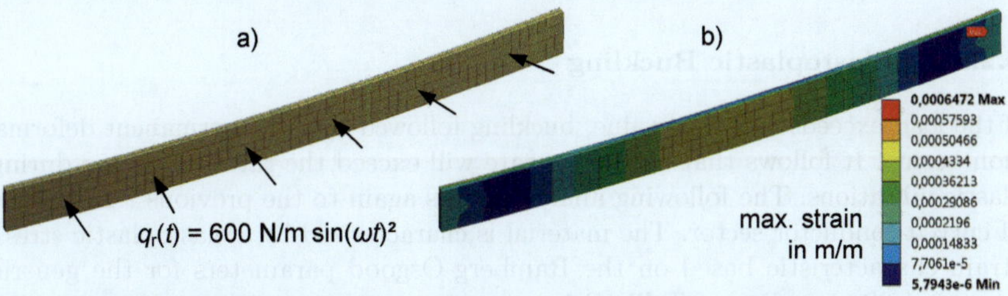

Figure 3.9: a) Undeformed FEA model with radial line load $q_r(t)$
b) Maximum strain at load amplitude of $600\,\text{N}\,\text{m}^{-1}$

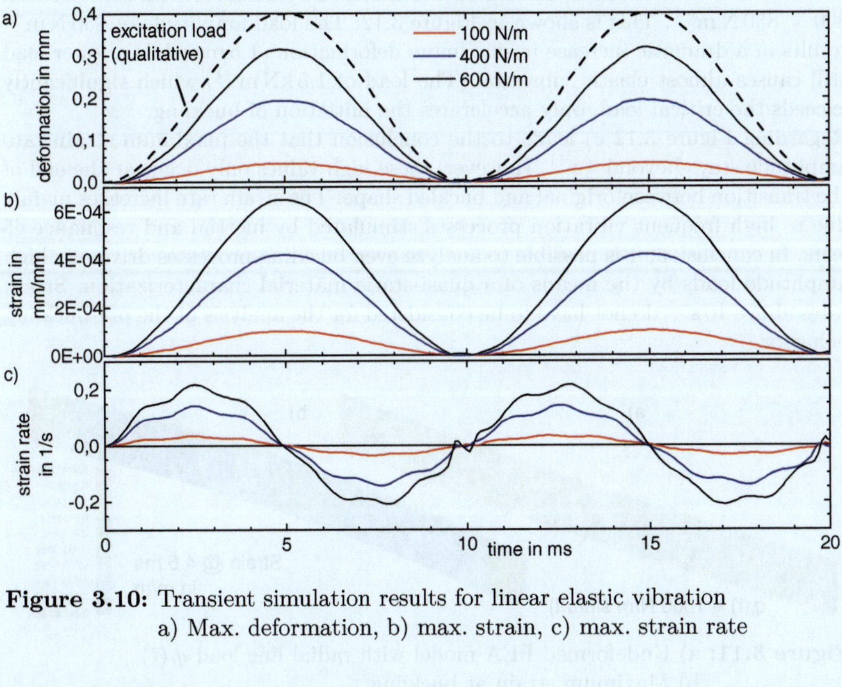

Figure 3.10: Transient simulation results for linear elastic vibration
a) Max. deformation, b) max. strain, c) max. strain rate

Table 3.4: Analytic results for the elastic vibration strain rate calculation

$\hat{q}_r$ in N/m	v_{max} in mm	σ_{max} in MPa	ϵ_{max}	δ_s in 1/s
100	0.043	11.2	$1.02 \cdot 10^{-4}$	0.032
400	0.172	44.8	$4.07 \cdot 10^{-4}$	0.128
600	0.258	67.1	$6.10 \cdot 10^{-4}$	0.192

3.2.2.2 Elastoplastic Buckling

If the load exceeds a critical value, buckling followed by large permanent deformation occurs. It follows that the strain rate will exceed the rates occurring during elastic vibrations. The following analysis refers again to the previous FEA model of curved conductor sector. The material is characterized by a elastoplastic stress-strain characteristic based on the Ramberg-Osgood parameters for the generic copper co-90 according to Table B.1.
Equation (3.28) applies to calculate the critical stress, as both conductor ends are fixed. Together with Equation (3.20) the static, critical load results in $q_{cr} = 1081.6\,\mathrm{N\,m^{-1}}$. The transient FEA model estimates a dynamic, critical load between $800\ldots850\,\mathrm{N\,m^{-1}}$. This is shown in Figure 3.12: The load amplitude of $850\,\mathrm{N\,m^{-1}}$ results in a dramatic increase in maximum deformation at 5 ms while a lower load still causes almost elastic vibration. The load of $1.5\,\mathrm{kN\,m^{-1}}$, which significantly exceeds the critical load, only accelerates the initiation of buckling.
Regarding Figure 3.12 c) leads to the conclusion that the maximum strain rate amplitude stays beyond $4\,\mathrm{s^{-1}}$. However, these high values only occur at the end of the transition between original and buckled shape: The strain rate increases mainly due to high frequent vibration processes stimulated by inertial and resonance effects. In conclusion, it is possible to analyze even buckling-processes driven by high amplitude loads by the means of a quasi-static material characterization. Strain rates above $10\,\mathrm{s^{-1}}$ do not have to be considered for the analysis of the pre-buckling behavior.

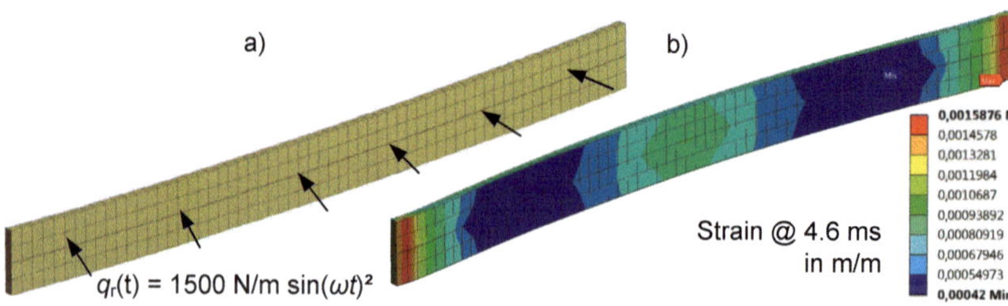

Figure 3.11: a) Undeformed FEA model with radial line load $q_r(t)$
b) Maximum strain at buckling

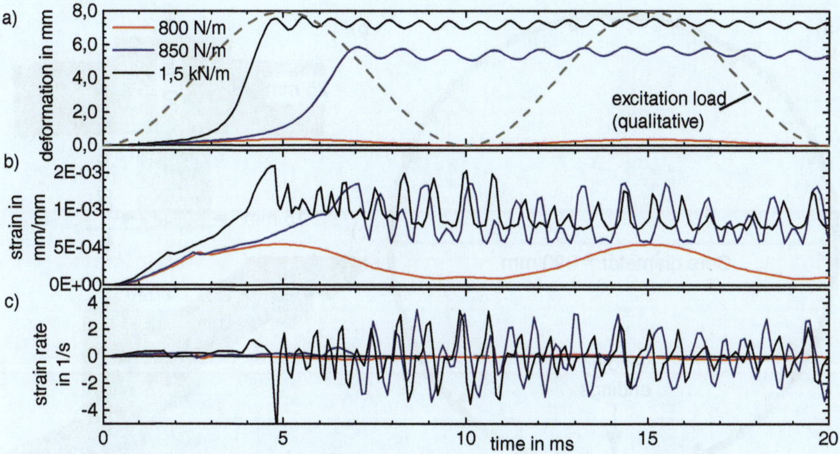

Figure 3.12: Transient simulation results for elastoplastic buckling: a) Max. deformation, b) max. strain, c) max. strain rate

3.2.3 Buckling of CTC Windings

3.2.3.1 Geometry and FEA Setup

This section analyzes the buckling theory described in Section 3.1.2 in more details by the means of FEA-based buckling studies. The aim is to validate the analytic approaches and point out limitations and influencing factors. Preliminary work has been done in [98, 92].

The underlying model geometry consists of a CTC turn supported by axial sticks and radial spacers as shown in Figure 3.13. The winding is not a closed ring but has the shape of a spiral. As the supporting iron core has very high stiffness, it is neglected. Radial spacers prevent axial conductor movements. The winding pitch corresponds to the sum of radial spacer height and CTC height. The width of the axial sticks is 15 mm, which corresponds to typical sticks used in power transformers. The radial width of 20 mm includes some of the elasticity of additional layers of Pressboard and insulating paper between core and winding.

The geometry parameters marked in Section 3.1.2 remain constant for all analyzes conducted within this section. The conductor is from type CTC-B-21 according to Table B.2. Strand edge radii and transpositions are neglected. The number of axial sticks N_{as} varies between 12 and 32 to consider different slenderness ratios. The Pressboard spacers have a linear elastic behavior based on a Young's Modulus of 1.2 GPa and a Poisson's ratio of 0.33. The copper characterization includes a nonlinear stress-strain characteristic with Ramberg-Osgood parameters according to co-90 and co-180 from Table B.1.

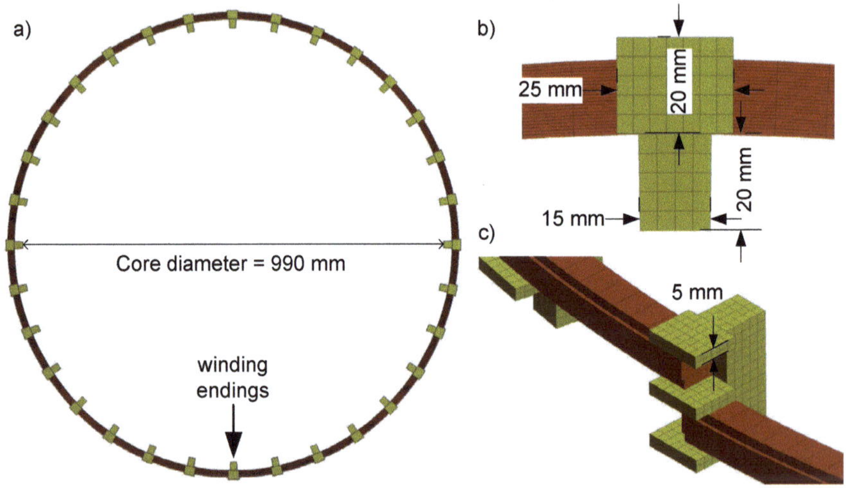

Figure 3.13: a) FEA model for buckling analysis,
b) top view on radial and axial spacers, c) winding endings

A regular mesh based on brick elements discretizes the model. The strand height and thickness directions are both discretized by two elements. In the direction of conductor length the element division is 300. The spacer discretization is visualized in Figure 3.13 b) and c).
Frictionless contacts define the numerous contacts arising between the strands itself or between strands and spacers. Bonded contacts fix the radial spacers to the axial sticks.
The inner surfaces of the axial sticks facing towards the core are fixed in all spatial directions. The winding endings shown in Figure 3.13 b) are also fixed. The buckling pre-force applies normal on the outer surface of each strand in the radial direction in the form of a perturbation pressure p_{p}. The final critical load results from the lowest load factor λ_1, perturbation pressure p_{p}, strand height h_{s} and number of strands N_{s} according to Equation (3.58).

$$q_{\mathrm{cr}} = \lambda_1 p_{\mathrm{p}} h_{\mathrm{s}} N_{\mathrm{s}} \tag{3.58}$$

3.2.3.2 Linear Analysis

The first step considers linear elastic copper properties, and the axial spacers can rotate with respect to the z-axis. This constraint leads to a winding support, which is close to the assumption of hinged supports made in the analytic considerations. Figure 3.14 visualizes the lowest three buckling modes for the model with 20 axial sticks. A factor of 76 scales the deformation. Note that the main result of a buckling

analysis are the load factor and its mode shape. The absolute deformation always refers to a unit deformation, which is 1 mm in this work.
The principle shape of all three modes is as expected from theory: One sector is radially bent inwards, while the adjacent is pushed outwards. However, the distribution of buckling waves is not equidistant. Some sections do not deform significantly.
Figure 3.16 a) compares the analytic critical loads according to Equation (3.19) to the lowest load resulting from linear FEA analysis. The coincidence is obvious, but the FEA results are slightly higher than the analytic values. One reason are the not perfect punctual hinged supports in the FEA analysis, and the inhomogeneity introduced by the rigid fixation of winding endings.

If a rigid support constraints the inner sides of the axial sticks, the critical loads increase dramatically, as the black squares of Figure 3.16 b) state. Regarding the lowest mode shapes in Figure 3.15 only a few sectors buckle. However, a scaling of analytic results to the simulative results is possible. This scaling factor is introduced within the slenderness ratio. It is named slenderness factor k_{sl} and increases the slenderness ratio according to

$$\tilde{T} = k_{\mathrm{sl}} \cdot T = k_{\mathrm{sl}} \cdot t_{\mathrm{s}} N_{\mathrm{as}} / D_{\mathrm{m}}. \tag{3.59}$$

Based on $k_{\mathrm{sl}} = 1.86$ the analytic results perfectly match the simulation results, as the blue and red data points in Figure 3.16 b) show.
The authors in [17, p. 275] propose a similar approach: They add a factor of 2 to the strand thickness t_{s}, in the case of buckling of unbonded CTCs. However, the present simulation results reveal, that this increase in buckling strength is not an effect of the CTC itself, but results from the stiffness of radial support structure.

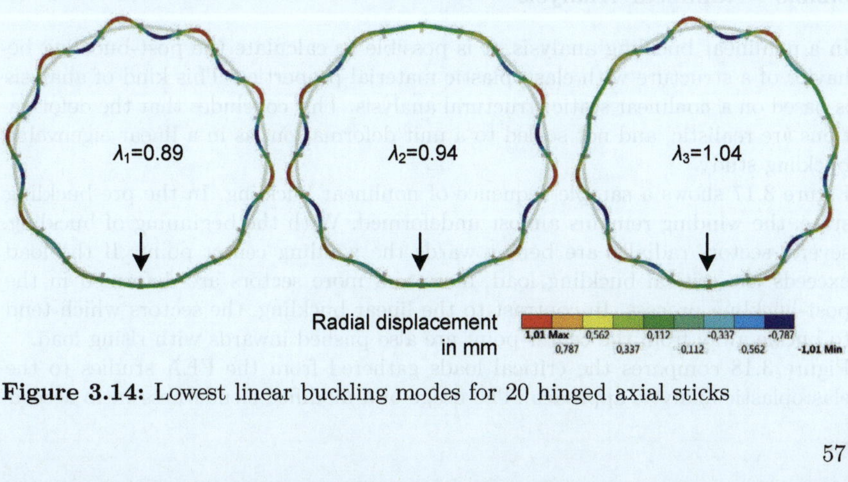

Figure 3.14: Lowest linear buckling modes for 20 hinged axial sticks

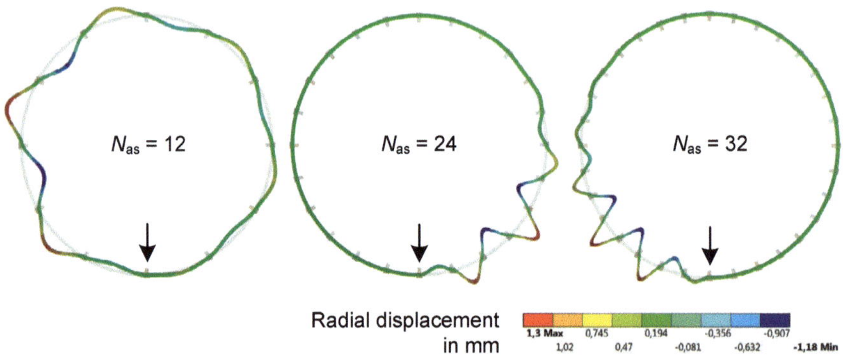

Figure 3.15: Lowest linear buckling modes for fixed axial sticks

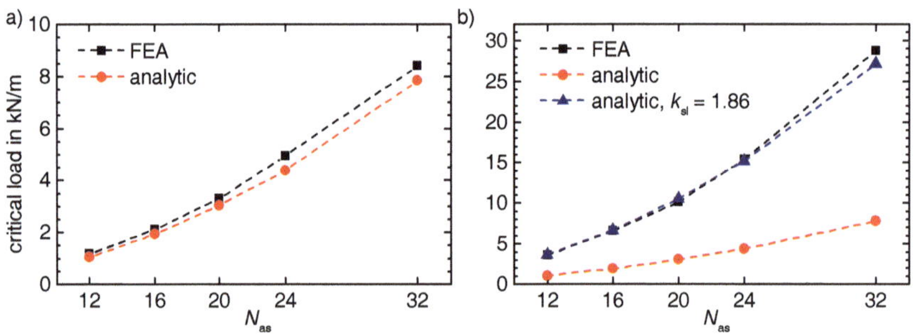

Figure 3.16: Comparison of linear FEA and analytic buckling analysis
a) Hinged axial sticks, b) fixed axial sticks

3.2.3.3 Nonlinear Analysis

In a nonlinear buckling analysis, it is possible to calculate the post-buckling behavior of a structure with elastoplastic material properties. This kind of analysis is based on a nonlinear static structural analysis. This concludes that the deformations are realistic, and not scaled to a unit deformation, as in a linear eigenvalue buckling study.
Figure 3.17 shows a sample sequence of nonlinear buckling. In the pre-buckling state, the winding remains almost undeformed. With the beginning of buckling, several sectors radially are bent towards the winding center point. If the load exceeds the critical buckling load, more and more sectors are deformed in the post-buckling process. In contrast to the linear buckling, the sectors which tend to buckle away from the center-point are also pushed inwards with rising load.
Figure 3.18 compares the critical loads gathered from the FEA studies to the elastoplastic analytic approach. The usage of a modified slenderness ratio accord-

ing to Equation (3.59), based on the factor $k_{sl} = 1.86$, again leads to a promising coincidence between FEA and analytic results.
For $N_{as} \in \{12, 16, 20\}$ the loads from the elastoplastic analysis are slightly higher than the loads gathered from linear FEA. This residual effect results from the different calculation methods of linear eigenvalue buckling and nonlinear buckling which is based on an elastoplastic structural analysis. A stress analysis for $N_{as} \in \{12, 16, 20\}$ reveals, that linear buckling is the predominant process, as no plastic strains are induced into the strands. Also, the analytic approach states a tangent modulus of 105.7 GPa for $N_{as} = 20$ with co-90 material parameterization, and 109.8 GPa with co-180 material. Both values are close to the Young's modulus of 110 GPa. Therefore, the winding already looses stability before the occurrence of plastic stress.
In the case of 32 axial sticks the tangent modulus is reduced to 75.2 GPa for co-90, and remains at 104.4 GPa for co-180. It follows that the first case is clearly plastic, while the latter case remains elastic.

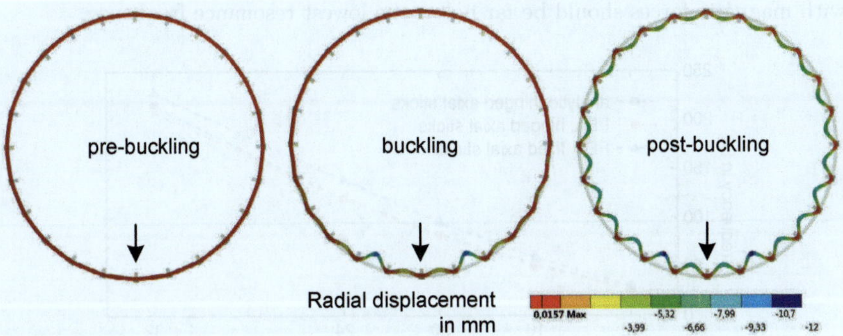

Figure 3.17: Development of nonlinear buckling with 24 fixed sticks

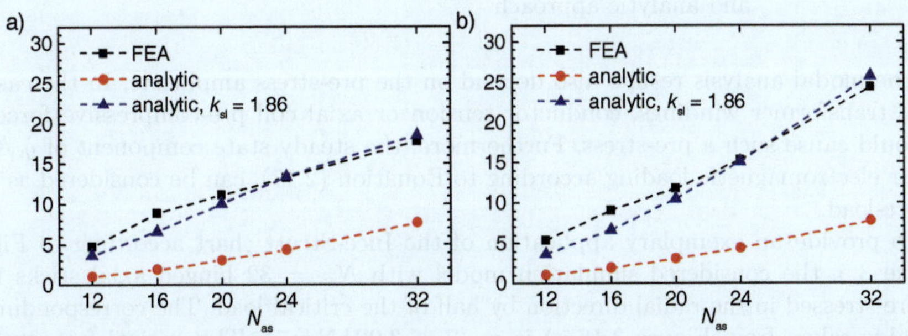

Figure 3.18: Comparison of nonlinear FEA and analytic elastoplastic buckling analysis. a) co-90, b) co-180

3.2.3.4 Modal Analysis

This dynamic FEA study analyzes a structural problem concerning possible eigenmodes, and solves for corresponding excitation frequencies. The eigenmodes refer to the linear eigenvalue buckling presented in Section 3.2.3.2. More details on the theory are given in [10].

The FEA geometry relates to the previous sections with consideration of both hinged and fixed supports at the axial sticks. Hydrodynamic mass per unit length k_{m} has been considered to be much smaller than the conductor mass per unit length.

Figure 3.19 compares the lowest critical frequencies gathered from the FEA modal analysis to the theoretical frequencies $f_{\mathrm{cr}} = \omega_{\mathrm{cr}}/(2\pi)$ with ω_{cr} according to Equation (3.39). For the case of hinged supports, the FEA results are close to the analytic frequencies. Rigid spacer fixation slightly increases the frequencies. This is a typical phenomenon also reported by other authors [10]: The resonance frequency increases with the system stiffness. Similarly, the number of axial sticks increases the critical frequencies. The main excitation frequency of 100 Hz arising with magnetic forces should be far below the lowest resonance frequency.

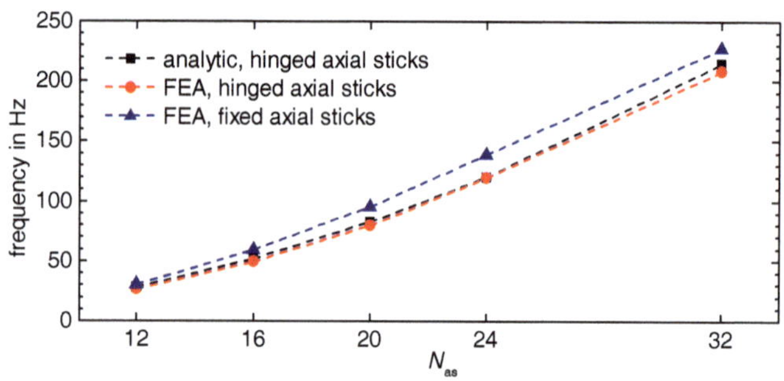

Figure 3.19: Critical frequencies for the first mode from FEA modal analysis and analytic approach

The modal analysis results also depend on the pre-stress amplitude. In the case of transformer windings, conductor tension or axial coil pre-compressive forces could cause such a pre-stress. Furthermore, the steady state component of $\hat{q}_{\mathrm{r}}/2$, for electromagnetic loading according to Equation (2.17), can be considered as a pre-load.

To provide an exemplary application of the Ince-Strutt chart according to Figure 3.5, the considered simulation model with $N_{\mathrm{as}} = 32$ hinged axial sticks is pre-stressed in the radial direction by half of the critical load. The corresponding value taken from Figure 3.16 a) is $q_{\mathrm{cr}}/2 = 3.92\,\mathrm{kN\,m^{-1}}$. The critical frequency for the first mode gathered from the simulation model reduces to $f_1 = 165.14\,\mathrm{Hz}$

as result of this pre-load. The critical frequency without pre-loading has been $f_{cr} = 214.59\,\mathrm{Hz}$, as visualized in Figure 3.19.
The corresponding Ince-Strutt stability parameters, calculated according to Equations (3.46) and (3.47), result in Equations (3.60) and (3.61). As it is of interest, whether the excitation frequency f_1 is unstable, the reference angular frequency in Equations (3.46) and (3.47) has to be $w_0 = 2\pi f_1$.

$$\nu = \left(\frac{f_{cr}}{f_1}\right)^2 \left(1 - \frac{q_{cr}/2}{2q_{cr}}\right) = 1.266 \tag{3.60}$$

$$\psi = \left(\frac{f_{cr}}{f_1}\right)^2 \frac{q_{cr}/2}{2q_{cr}} = 0.422 \tag{3.61}$$

This parameter-set is drawn in the Ince-Strutt diagram of Figure 3.5 by the means of a black cross. It is located in the stable, gray area, but very close to a stability boundary. Taking into account, that the numerical analysis implies some minor imperfections (like the not perfectly hinged supports), this result is satisfying, as the point is very close to the instable, white areas of the Ince-Strutt diagram.

This concludes that with increasing load, the span of critical frequencies increases, as the red parameter lines in Figure 3.5 cut more white regions. For decreasing loads, the critical frequency approaches $\nu = 1$, which leads according to Equation (3.46) to $\omega_0 = \omega_{cr}$. The critical frequency is defined according to Equation (3.39).

Chapter 4

Static Characterization of Conductors

4.1 Testing Methods and Standards

4.1.1 Tensile Test

Tensile testing is one of the most important methods to characterize the elasticity of materials. This test provides the material's response to unidirectional stress in the form of a stress-strain relationship. The principal test arrangement is to subject a sample to controlled tension until failure. Regarding transformer conductors, the following three tensile testing standards are relevant:

DIN EN ISO 6892-1:2009 [74] defines the tensile testing for metals, and is therefore suitable for the copper strands. The test specimen shape is standardized by DIN 50125:2009 [67]: Two shoulders with a gauge section in between, for both round or flat test samples. The gauge cross-section has to be much smaller than the shoulders, so that the main deformation and failure will occur in this area. Due to the slender shape of CTC strands, it is not possible to prepare fully standard-compliant test specimens. The outer strand dimensions limit both thickness and width of the shoulders. To reduce the influence of mechanical machining, only the gauge-section should be cut out by a milling-machine as shown in Figure 4.1 a). The remaining dimensions should correspond to the original strand dimensions.
An extensometer applied to the gauge section during test enables precise strain measurements. Calculating the strain from the traverse position of the testing apparatus is defective, as the shoulders' elongation and the testing apparatus stiffness influence the results.
Dividing the measured force through the cross-sectional area of the gauge section leads to the engineering stress σ_e. Furthermore, the elongation measured by the extensometer provides the engineering strain ϵ_e regarding the original gauge section length. However, a precise material parameterization for FEA software often demands for the true values:

The true strain ϵ_t refers to the instantaneous increase in the instantaneous gauge section length, and is calculated from engineering strain according to Equation (4.1) [32, p. 24]. The true stress σ_t takes cross-sectional contraction for large strains into account, as defined by Equation (4.2)[32, p. 24].

$$\epsilon_t = \ln(1 + \epsilon_e) \tag{4.1}$$

$$\sigma_t = \sigma_e(1 + \epsilon_e) \tag{4.2}$$

The zero-span tensile test is suitable to characterize the insulating paper by the means of its fiber strength. Correlated standards are ASTM-D5804-97:2002 [70], ISO 15361:2000 [76] and TAPPI T231:1996 [79]. This test demands for special clamping jaws, which facilitate a precise clamping of the paper strips close to the outer edges of the jaws, as shown in Figure 4.1 b). The initial clamping length should be below 100 µm, to ensure that only the fibers itself are stressed and not the inter-fiber bondings. Unless otherwise mentioned, the test specimens are stretched uniformly with a constant strain rate of $20\,\mathrm{mm\,min^{-1}}$ until break. The test sample's width equals the strip width, as long it does not exceed the clamping jaws' width. The ultimate tensile strength deduced from this test characterizes the fiber strength of the paper. It is possible to correlate this mechanical property to the chemical pendant degree of polymerization (DP), which also describes the fiber strength of cellulose [25, 93]. The stress-strain curve has no significance, as a precise strain measurement due to the small clamping length is not possible.

The wide-span tensile test is suitable to characterize the strength of inter-fiber bondings of insulating paper. The underlying standard used in this work is DIN EN ISO1924-2:2009 [73]. This method provides a stress-strain characteristic similar to the tensile test for metals, which is suitable for FEA-based or analytic strength analysis.

Although the span distance between the clamping jaws has to be 180 mm according to the standard, a reduced span length of 20 mm is used in this work, as visualized in Figure 4.1 c). This relates to the case of a paper wrapped conductor, where only a few centimeters of strip length between two conductor edges are stressed if the conductor bends. Furthermore, testing with reduced span distance saves material, which is mainly necessary for tests on aged samples. Due to the limited strip width of usually $< 30\,\mathrm{mm}$ it is very difficult to clamp and measure CD test specimen properly. Therefore, the results presented in this work are based on the MD properties. Unless otherwise stated, the strain rate is $20\,\mathrm{mm\,min^{-1}}$ according to the standard. A change in strain rate influences the tensile strength, as cellulose exhibits both elastic and flow properties. At high strain rates the elastic properties dominate, resulting in a higher tensile strength. At low strain rates the paper exhibits flow making the paper break at lower loads, as the fibers have more time to slide against each other [25]. For the zero-span measurement, this effect is not that relevant, as the inter-fiber bondings are not stressed.

A precise strain measurement through the traverse position demands a compensation of testing machine modulus. For the used testing apparatus zwicki-line Z2.5TN from Zwick GmbH & Co. KG the machine modulus has been identified at 2.51 N μm^{-1}, by the use of a massive steel block as test sample. In this case, the measured strain mainly results from the deflection of the testing machine itself.

4.1.2 Three-Point Bending Test

The three-point bending test characterizes the bending stiffness of a conductor. Underlying standards used in this work are DIN EN ISO 7438:2012 [75], which is suitable for metals and DIN EN ISO 178:2013 [77] for plastics.
Circular shaped supports hold the conductor, while a loading stamp lowers from above, as shown in Figure 4.2. This test stresses the conductor similarly to a radial compressive electromagnetic load acting on a winding sector between two spacers. Unless otherwise stated, the typical distance between the lower supports is 180 mm, and the stamp displacement rate is 10 mm min^{-1}. Both stamp and supports are of cylindrical shape with a radius of 10 mm. To suppress the influence of shearing stresses, the span to specimen-height ratio has to be larger than 16 [77].
A 2.5 kN load cell mounted to the stamp measures the applied force. The displacement is derived from the traverse position, with compensation of machine modulus as explained in the previous section. For the present test setup, the machine modulus is 2.34 N μm^{-1}. Some exemplary measurement results are given in Figure 4.5.

4.1.3 T-Peel Test

This test characterizes the peel resistance of the adhesive coating used within bonded CTCs. The underlying standard DIN EN ISO 11339:2010 [72] defines a bonding T-peel test for flexible-to-flexible bonded assemblies, by the means of a T-type specimen.
In deviation from the standard, the test specimen dimension is limited by the size of the CTC strands. It has not been possible to obtain the adhesive separately and apply it to other test samples. Therefore, test specimen have been made from two strands, bent by a right angle, as shown in Figure 4.3 a). The bonded area length is 95 mm. For thermal activation of the adhesive, the CTC manufacturer demands a temperature of 120 °C for 24 h and a pressure on the bonded area of 1 MPa.
The span length of the tensile testing apparatus is 100 mm and the strain rate equals 4 mm s^{-1}. The test result is the force-displacement characteristic until the bonded area is completely delaminated. The traverse position with compensation of machine modulus provides the displacement information. Some exemplary measurement results are given in Figure 4.25.

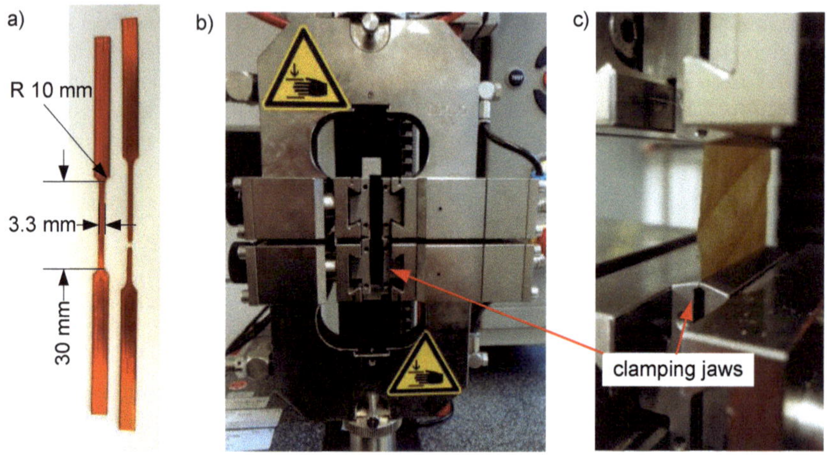

Figure 4.1: a) CTC strand test specimen before and after tensile test
b) Zero-span measurement setup with clamping jaws opened
c) Clamped wide-span insulating paper test sample

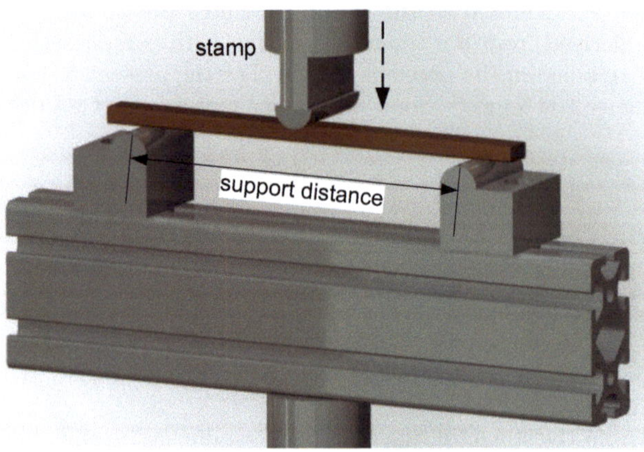

Figure 4.2: Three-point bending test arrangement

4.1.4 Overlap Shear Test

The overlap shear test measures the shear strength of the adhesive used within bonded CTCs. The testing method is based on ASTM D1002-10 [69]. The underlying test specimen are made from two rectangular sections, which are bonded on an overlap length as visualized in Figure 4.3 c). Once again, in deviation from the standard, the test specimen dimension is limited by the CTC strands. Especially in this test, the carrier object stiffness has an impact on the results. However, the testing material has only been available in the form of coated CTC strands, and a separate application of the adhesive to other carrier objects has not been possible. The thermal activation process is the same as described in Section 4.1.3. The clamping length of is 145 mm, and the strain rate equals 50 mm min^{-1}. A relocation of clamping jaws by the strand thickness t_s prevents eccentric loading. The test result is the force-displacement characteristic until the bonded area is completely delaminated. The traverse position with a compensation of the machine modulus provides the displacement information. Some exemplary measurement results are given in Figure 4.26.

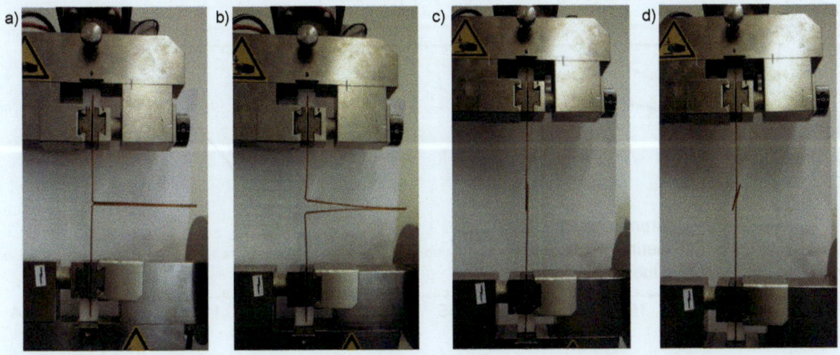

Figure 4.3: a) T-Peel test setup, b) Delaminated T-peel test specimen c) Shear test setup, d) Partially delaminated shear test specimen

4.2 Preliminary Investigations

The technical data of all CTCs used within this work is listed in Table B.2. The labeling contains the number of strands to simplify the assignment. The parameter p describes the nominal (measured) increase of the outer CTC dimensions due to the insulating paper. This value is defined by the paper thickness d_p and number of layers N_p to $p \approx 2N_p d_p$.

4.2.1 Copper Characterization

Figure 4.4 a) visualizes six exemplary tensile test results conducted with strands from CTC-B-21. Especially in the moment of failure the strain shows large scattering, but also the remaining curves differ slightly from each other. However, for the simulations conducted in this work, it is sufficient to focus on the stress-strain characteristic far below the strain at break.

Therefore, the red line provides the average engineering stress up to an engineering strain of 20 %. The calculated true stress-strain curve (blue dotted line) runs above the engineering values, due to the stress increase resulting from the cross-sectional contraction. The expanded view in Figure 4.4 b) proves that engineering and true values are identical for small strains. The copper Young's modulus E_0 results from the gradient inside the linear elastic range (black dotted line).
Table 4.1 lists the Young's modulus and elastoplastic Ramberg-Osgood parameters according to Section 2.3.3. This table only considers conductors, for which a precise elastoplastic characterization is necessary in the subsequent investigations. The Young's modulus is very close to the literature value of 110 GPa for annealed copper, and also the yield strengths are quite similarly. In conclusion, the four CTCs are made from nearly the same copper alloy.

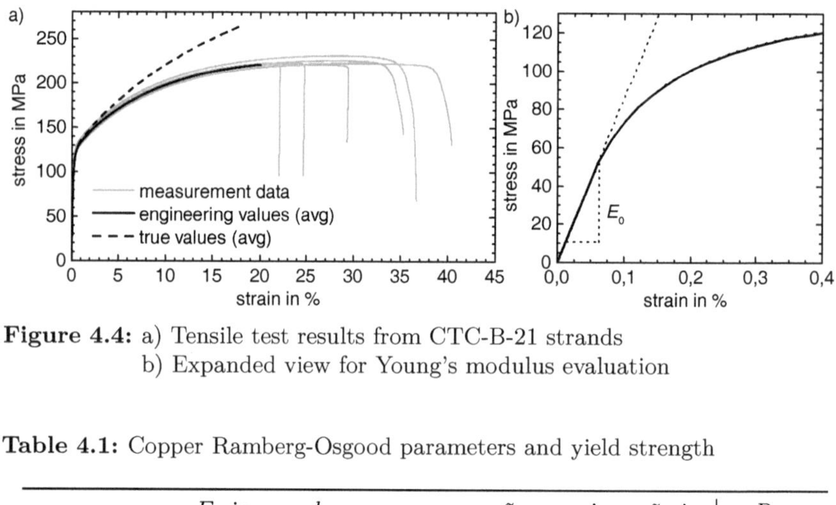

Figure 4.4: a) Tensile test results from CTC-B-21 strands
b) Expanded view for Young's modulus evaluation

Table 4.1: Copper Ramberg-Osgood parameters and yield strength

	E_0 in GPa	k_R	m	$\tilde{m}$	σ_0 in MPa	$\tilde{\sigma}_0$ in MPa	$R_{p,0.2}$ in MPa
CTC-A-11	95.6	3/7	4.28	1.61	63.7	0.81	100.4
CTC-B-21	95.6	3/7	5.54	1.71	86.1	1.73	116.8
CTC-D-37	95.6	3/7	3.95	1.74	70.0	1.97	111.9
CTC-E-19	95.9	3/7	7.38	1.60	94.4	1.18	116.5

4.2.2 Validation of Simplified CTC Model

To validate the simplified CTC model presented in Section 3.2.1, a bare CTC without insulating paper has been compared to a stack of straight strands by the

means of bending tests. The stack has been piled up by hand based on straight strands taken from the same CTC type. In conclusion, it has no transpositions. According to the theory of Equation (3.4), the stiffness ratio between both tests has to be equal to the ratio between the number of strands for the linear elastic load case. Therefore, a scaling of the stack force-deflection curve by this ratio should correspond to the CTC force-displacement curve.
The first conductor type CTC-A-11 has been measured at a long support distance of 180 mm as a worst-case consideration. The ratio between number of transposing strands and total number of strands is 4/11, and the vast support distance extends over five transpositions: Three on the top side, and two on the bottom side of the conductor. The applied oil impregnation reduces frictional effects from the unbonded thermosetting coating.
Another CTC from type CTC-B-21 has been measured with a short support distance of 100 mm. The ratio 4/21 between transposing strands to total number of strands is smaller compared to CTC-A-11, and only two transpositions lie in-between the supports. In this test the transposing strands should have a significantly lower impact on the stiffness compared to the first conductor.
For both tests the scaled stack results perfectly match the CTC results in the linear elastic range, as Figure 4.5 proves. For deflections exceeding the linear range, the CTCs are slightly stiffer. To conclude, the applicability of the simplified CTC model, which already has been shown by the means of simulations, also results from the bending tests.

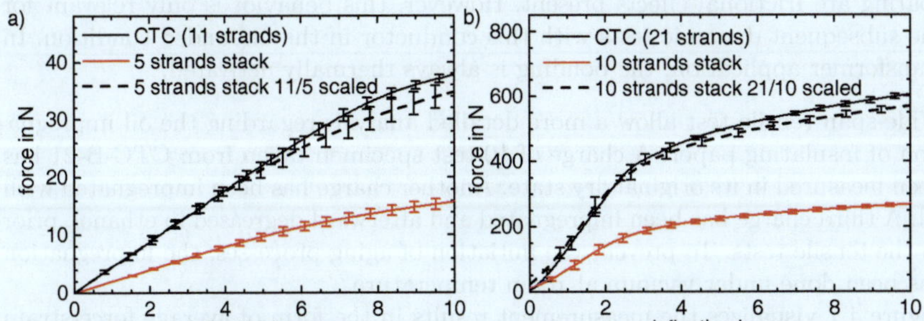

Figure 4.5: Three-point bending tests comparing bare CTCs to straight stacks
a) CTC-A-11 unbonded, impregnated, support distance: 180 mm
b) CTC-B-21 impregnated, support distance: 100 mm

4.2.3 Oil Impregnation Effects

Besides enhancing the insulating system's dielectric strength, the oil used in power transformers has also impact on the mechanical properties of conductors: It reduces frictional effects between the strands of a CTC. The impact of oil on the elastoplastic properties of insulating paper is another uncertain topic, which has not been investigated yet. As the dynamic tests presented in Chapter 5 are conducted in the dry state, these pre-investigations are mandatory.

The comparison of dry and impregnated factory new CTC samples by three-point bending tests facilitates an integral consideration to the impact of frictional effects. The same tests conducted with stripped CTCs evaluate the impact of friction between the copper strands.
Figure 4.6 shows measurement results for CTC-B-21 and CTC-D-37 in the form of averaged force-displacement curves. The error bars refer to the standard deviation. The CTC-D-37 remains unbonded, which causes a significant higher frictional coefficient between the strands due to the adhesive coating, in contrast to CTC-B-21. The results from CTC-B-21 show no significant difference in bending behavior between the impregnated and the dry state, regardless of the presence of paper insulation.
The dry, stripped CTC-D-37, however, has a stiffer characteristic than its impregnated counterpart. This is due to the unbonded adhesive layer, with its high frictional coefficient, which transfers shear between the strands. However, this contribution is not relevant in the presence of a paper insulation: Both curves from impregnated and dry measurements are identical. Impregnation only reduces the variance: The bigger error bars correspond to the dry samples.
The overall conclusion is that the reduction in frictional coefficients through transformer oil plays no significant role. Only in the case of a bonded CTC with unbonded coating are frictional effects present. However, this behavior is only relevant for the subsequent dynamic tests with this conductor in the unbonded condition. In transformer application, the bonding is always thermally activated.

Wide-span tensile test allow a more detailed analysis regarding the oil impregnation of insulating paper. A charge of 10 test specimen taken from CTC-B-21 has been measured in its original dry state. Another charge has been impregnated with oil. A third charge has been impregnated and afterward degreased in ethanol, prior to the tensile tests. To prevent the initiation of aging processes, the impregnation has been done under vacuum at room temperature.
Figure 4.7 visualizes the measurement results in the form of average force-strain characteristic with error bars representing the standard deviation. A pre-load of 10 N has been chosen as starting point to plane the samples, which had some minor kinks and creases due to the wrapping around the CTC.
It is evident that the ultimate strain decreases from 4.8 % to 3.4 %, as an effect of impregnation. A subsequent degrease does not reverse this phenomenon. Furthermore, the scattering irreversibly decreases due to the impregnation. Both the impregnated and degreased samples behave quite identically. The maximum tensile force of around 216 N is the same for all three charges.
In conclusion, the oil impregnation seems to straighten the paper strips, and leads to some kind of shrinkage, as the maximum strain at failure is reduced. This phenomenon has also been reported previously for Pressboard in [45, p. 76-77] and

[53]. As both impregnated and degreased samples behave identically this effect is irreversible, and does not depend only on the oil itself. Referring to the measurements of Figure 4.6, this shrinkage effect has no significant impact on the bending behavior of the CTCs. Only the variance decreases. Another conclusion is that for the zero- or wide-span measurements on aged and impregnated paper strips a degreasing process is not necessary.

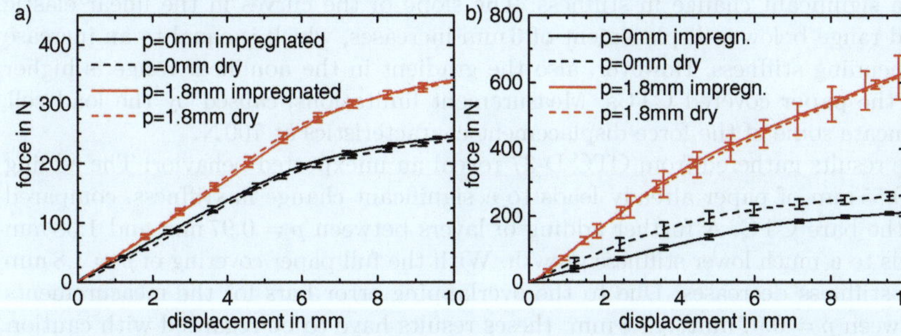

Figure 4.6: Impact of oil impregnation on three-point bending test results with factory new CTCs: a) CTC-B-21, b) CTC-D-37 unbonded

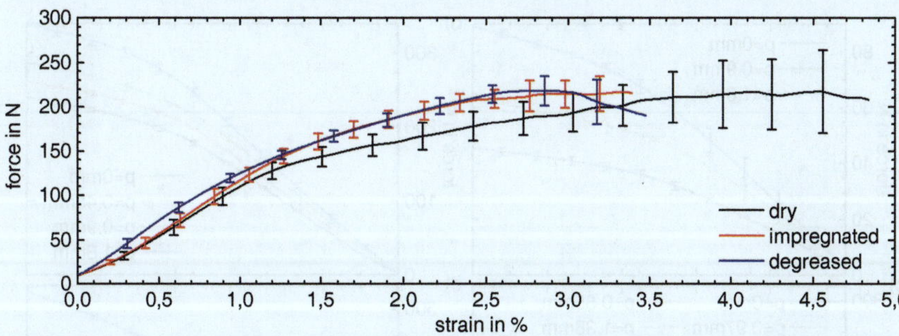

Figure 4.7: Force-strain characteristics deduced from wide-span tensile tests at dry, impregnated and degreased factory new insulating paper (MD)

4.3 Stiffness Contribution of Insulating Paper

4.3.1 Measurement Results

The results of Figure 4.6 already unveiled a significant impact of paper insulation thickness p on the flexural characteristic of a CTC: The slope in the linear elastic range increases due to the paper covering.

This section provides a more detailed analysis based on a variation in p, by unwrapping a specific number of layers. Figure 4.8 visualizes the corresponding bending test results for several CTC types. The average curves and error bars are based on three tests, except CTC-C-29 whose data is based on two measurements with and one measurement without paper. Due to limitations in the availability of testing material, the degree of variation in p is different for the considered CTCs.
All measurements share one commonality: The presence of insulating paper leads to a significant change in stiffness: The slope of the curves in the linear elastic load range below a displacement of 6 mm increases, which is equal to an increase in bending stiffness. However, also the gradient in the nonlinear range is higher for the paper covered CTCs. Measurement limitations, caused by the load cell, truncate some of the force-displacement characteristics at 700 N.
The results gathered from CTC-D-37 reveal an unexpected behavior: The adding of 0.55 mm of paper already leads to a significant change in stiffness, compared to the bare CTC. A further adding of layers between $p = 0.97$ mm and 1.38 mm leads to a much lower stiffness growth. With the full paper covering of $p = 1.8$ mm the stiffness decreases. Due to the overlapping error bars for the measurements between $p = 0.97$ mm ... 1.8 mm, theses results have to be regarded with caution. However, a saturation effect seems to exist concerning the paper insulation stiffness contribution with increasing number of layers.

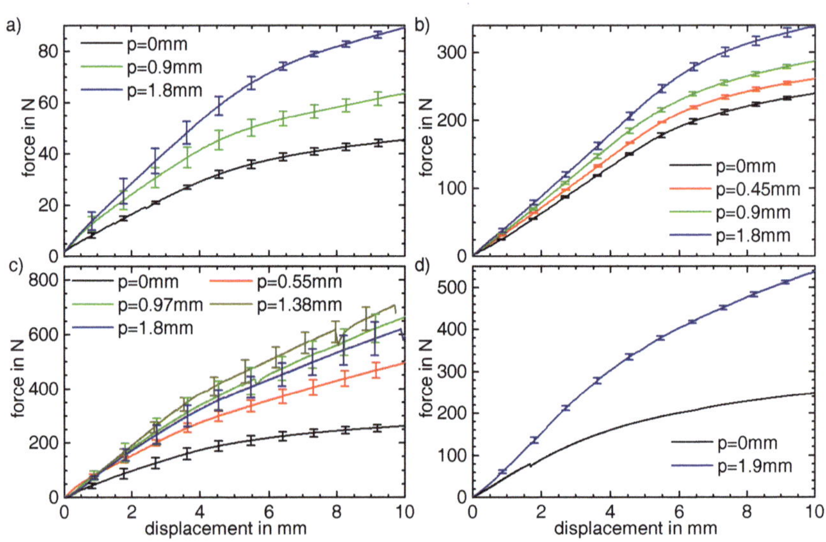

Figure 4.8: Force-displacement characteristics deduced from three-point bending tests at factory new CTCs with different paper increase p (dry)
a) CTC-A-11, b) CTC-B-21, c) CTC-D-37, d) CTC-C-29

4.3.2 Equivalent Stiffness Evaluation

This section describes the contribution of the paper insulation to the total conductor bending stiffness by the means of an equivalent stiffness. The slope of the linear elastic range at the beginning of three-point bending test results characterizes the bending stiffness: According to [11] the deflection x for a simply supported straight beam with support distance L, loaded by a central force F is

$$x = \frac{FL^3}{48E_0I}. \tag{4.3}$$

Rewritten in the form of Equation (4.4) gives the bending stiffness S depending on support distance L and the ratio F/x. The latter corresponds to the slope of the force-deflection characteristic in the linear elastic load range.

$$S = E_0I = \frac{F}{x} \cdot \frac{L^3}{48} \tag{4.4}$$

Two bending tests are necessary to estimate the paper insulation stiffness contribution: One with a stripped CTC resulting in the copper stiffness contribution S_{co}, and the other with a paper covered CTC of the same type providing S_{p}. The difference between both is the stiffness of paper covering according to

$$S_{\mathrm{pap}} = S_{\mathrm{p}} - S_{\mathrm{co}}. \tag{4.5}$$

This approach bases on the assumption that both structural elements copper core and paper covering experience a coherent deformation. The derivation of the underlying equations is given in Appendix A.1. Based on the equivalent stiffness for the overall paper covering, it is possible to calculate a homogenized Young's modulus as follows:

$$E_{\mathrm{pap}} = \frac{S_{\mathrm{pap}}}{I_{\mathrm{pap}}} = \frac{S_{\mathrm{p}} - S_{\mathrm{co}}}{I_{\mathrm{pap}}} = \frac{S_{\mathrm{p}} - S_{\mathrm{co}}}{(h_{\mathrm{c}} + p)(t_{\mathrm{c}} + p)^3/12 - h_{\mathrm{c}}t_{\mathrm{c}}^3/12} \tag{4.6}$$

This equation includes the area moment of inertia for the paper covering I_{pap}, which concludes from the formula of a rectangular shaped box girder with the outer dimensions of the inner copper core h_{c}, t_{c}. More details to the calculation are given in [88].

Figure 4.9 visualizes the results gathered from the measurement data of Figure 4.8. For CTC-B-21 and CTC-C-37 additional measurements have been carried out in the impregnated condition.
The percentage stiffness increase due to the paper insulation is calculated to $(S_{\mathrm{p}} - S_{\mathrm{co}})/S_{\mathrm{co}}$. This ratio not only depends on the total paper covering thickness, but also on the CTC cross-sectional parameters.

In the case of CTC-D-37 there also is a considerable difference between dry and impregnated tests for $p = 1.8\,\text{mm}$, resulting from the unbonded adhesive coating. The impregnation reduces the frictional coefficient between the strands, as S_{co} is also lower. As a result, the paper stiffness contribution increases. This difference does not occur for CTC-B-21 without adhesive coating.

The stiffness decrease for CTC-D-37 with increasing paper thickness from $p = 0.97\,\text{mm}$ to $p = 1.80\,\text{mm}$ has already been mentioned in the previous section. This behavior shows a saturation effect concerning the stiffening properties of the paper insulation. On the one hand side more layers stiffen the paper covering, but as a consequence the shearing stresses and correlated shearing deformation also increase. It is, therefore, CTC-D-37 with the largest radial thickness, which shows this effect.

The homogenized Young's modulus not only depends on the total paper covering thickness p, but also on the conductor geometry itself. Shearing stresses, which increase with rising radial conductor dimension and total paper thickness affect the calculation of E_{pap}, and are very difficult to describe.
The difference between impregnated and dry condition does not affect the Young's modulus that much compared to the stiffness increase. The results for CTC-B-21 are almost equal. In the case of CTC-D-37 the dry test results in 0.87 GPa, while the impregnated test leads to a Young's modulus of 1.22 GPa.
However, even the maximum homogenized Young's modulus of 2 GPa is much lower than the paper strip Young's modulus of 5 GPa, which has been derived from wide-span tensile tests (Table 4.2). The homogenized Young's modulus has to be lower, due to the wrapped construction of the paper insulation, which behaves not as stiff as the assumption of a homogenized cross-section. Furthermore, the anisotropic properties of cellulose contribute to this effect. The Young's modulus and tensile strength σ_{t} in the cross direction usually are lower compared to the manufacturing direction. The author of [62] lists values of $E_0 = 2.5\,\text{GPa}$, $\sigma_{\text{u}} = 33\,\text{MPa}$ in CD and $E_0 = 3.5\,\text{GPa}$, $\sigma_{\text{u}} = 57\,\text{MPa}$ in MD.

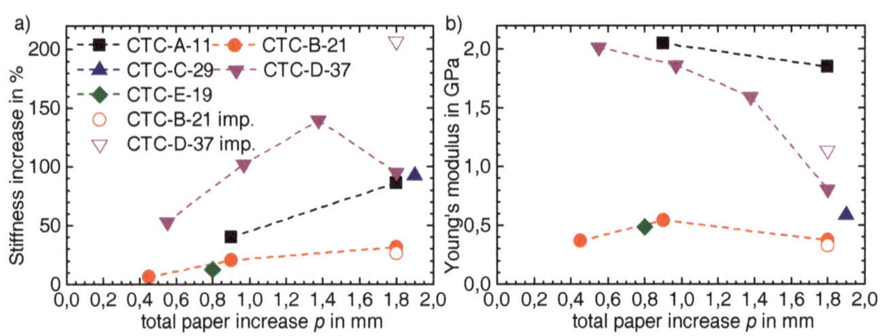

Figure 4.9: a) Percentage stiffness increase
b) Homogenized Young's modulus of paper covering

4.4 Impact of Insulating Paper Aging

Artificially aged CTCs have been analyzed by the means of bending tests to evaluate the effects of insulating paper aging on the bending stiffness. Additional zero- and wide-span tensile tests and DP analysis characterize the aging state of the insulating paper. Preliminary investigations to the tensile tests are given with the student thesis [93]. Finally, an FEA-based model with aging dependent material parameters enables a deeper analysis of the failure processes.

4.4.1 Accelerated Aging Procedure

With accelerated aging it is possible to age the insulation of CTCs within a reasonable time. The test samples are 27 curved pieces of CTC-B-21 with a length of 220 mm and a curvature radius of 430 mm. The latter results from the coiling radius of the reel. Figure 4.10 a) is a schematic representation of a test specimen placed in the three-point bending test.
The CTC pieces have been stored inside a stainless steel container (40 cm × 60 cm × 25 cm), filled with 30 l of factory new Shell Diala DX oil, as shown in Figure 4.10 b) and c). The samples have not been dried before aging, to speed up the aging process due to the presence of water in the cellulose. The insulating papers moisture content was 8 %. The steel container has been placed inside a vacuum oven at a temperature of 115 °C ± 1.5 °C. At the start of the aging procedure and after each sample recovery this oven has been evacuated and flooded with argon. This method impedes further aging caused by moisture and oxygen from the ambient air. The aging periods have been:

6, 13, 25, 35, 59, 90, 162 days

The reference measurement on factory new samples is named "0 days" in the following.

4.4.2 Bending Test Results

For each aging state 4 test specimens have been recovered from the oven. Figure 4.11 visualizes the averaged force-displacement characteristics, measured at room temperature. To enhance clearness error bars showing the standard deviation are only given for the measurement results in Figure 4.12.

The force-displacement characteristics gathered from aged samples start with a significant higher slope, compared to the factory new specimens. This steep increase is followed by a maximum and a scattered decrease, until all measurement results coincidence for displacements between 7 . . . 12 mm. The height of the initial maximum grows with aging time, starting at 100 N for the 6 days measurement, and ending up with 300 N, after 162 days.

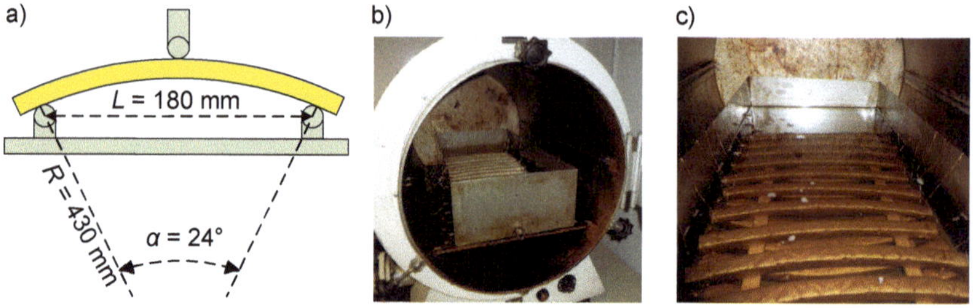

Figure 4.10: a) Geometry of test specimens, b) vacuum oven with container, c) arrangement of CTC pieces in oil

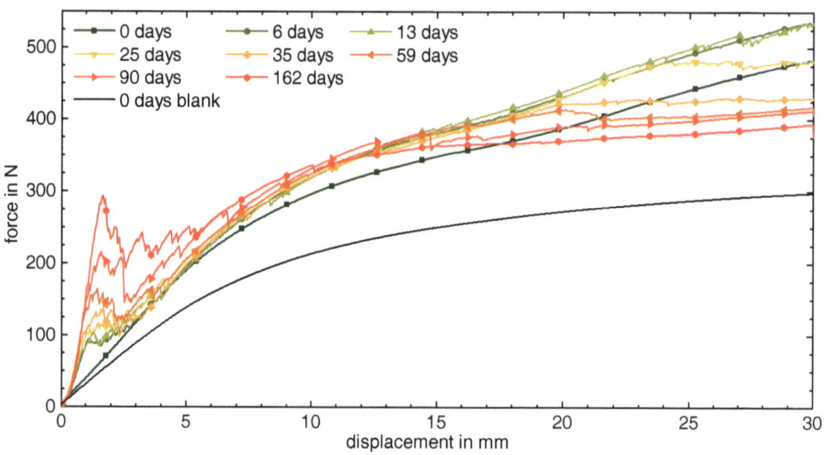

Figure 4.11: Three-point bending results for aged CTC samples

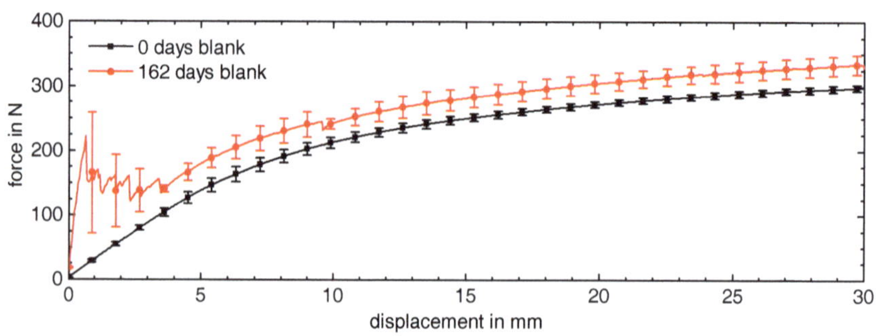

Figure 4.12: Three-point bending results for blank CTCs

In order to analyze the reason for this stiffness increase, the paper has been removed from some 162 days old samples prior to the bending test. There was no aging dependent adhesion between the layers. However, the bending test results with these bare, aged CTCs show the same behavior like the paper covered aged samples for displacements below 5 mm (Figure 4.12). Only the initial slope is slightly lower for the paper covered measurements, as the initial compression of the paper covering close to the supports and the stamp induces residual displacements.
To conclude, a weak sticking between the strands develops with proceeding aging time due to chemical reaction products, resulting from oil and cellulose decomposition. Due to the sticking effect it is not possible to evaluate the stiffness contribution of aged insulating paper from the bending tests, as it has been done in Section 4.3.2. Furthermore, the sticking impact slightly is present even for large deflections, as there still is an offset between the curves shown in Figure 4.12.

For displacements above 5 mm, the aging dependent curves of Figure 4.11 start to flatten. The associated critical deflection decreases with proceeding aging time. The gradient of the force-displacement curves for deformations above these points approaches the gradient of the blank conductor measurement result. The first derivative of the force displacement curves estimates these critical points more clearly and overcomes the offset problem due to the sticking. The resulting curves approximately are proportional to the stiffness, as Equation (4.4) for a straight beam states. The vertical dashed lines in Figure 4.13 mark the deflection, where the derivatives of the aged measurements fall below the black curve for the blank CTC. These points define the mechanical breakdown of the insulating paper. Furthermore, the paper insulation stiffness contribution is almost lost. A visual observation of the test samples during the bending test unveiled distinct cracking in the middle of the test samples from this point on (Figure 4.14). Furthermore, the unwrapped damaged layers prove that even the innermost layers are broken (Figure 4.15).

4.4.3 Insulating Paper Characterization

The mechanical properties of the insulating paper experience severe change during the aging process. However, aging is not distributed uniformly throughout the layers. It follows the thermal, moisture and oxidation gradients. The change in color proves this assumption, as the outer layers are much darker than the inner ones (Figure 4.15). Aging products dissolving in oil, like moisture or acids, can affect the outer layers much better, due to the better contact to the oil.
In contrast, aged conductors taken from a transformer show significant aging for the innermost layers. As these layers are in direct contact to the heat generating conductor they suffer higher temperature, which speeds up thermal aging [33, 40]. The present accelerated aging method, however, is in thermal steady state.
For each conductor sample the four inner and outer layers have been analyzed using the zero- and wide-span tensile testing method in MD direction. A pre-force of 10 N removes residual strain resulting from a flattening of kinks and small wrinkles in the paper strips. For the heavily aged samples older than 59 days a reduction to

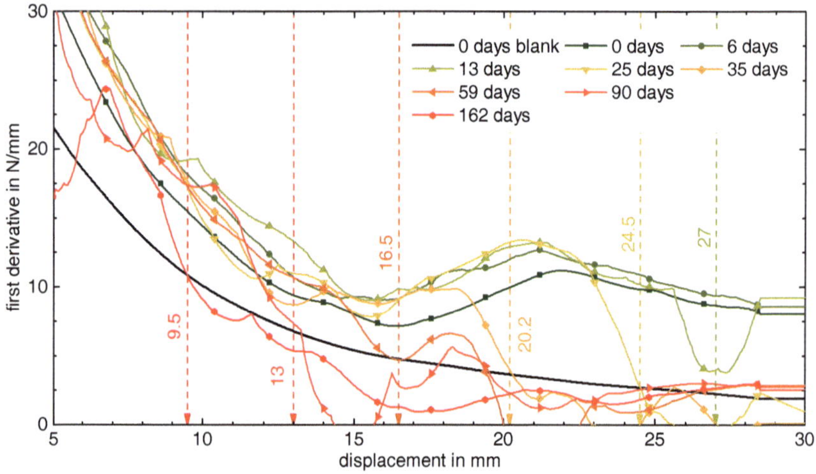

Figure 4.13: First derivative of force-displacement curves

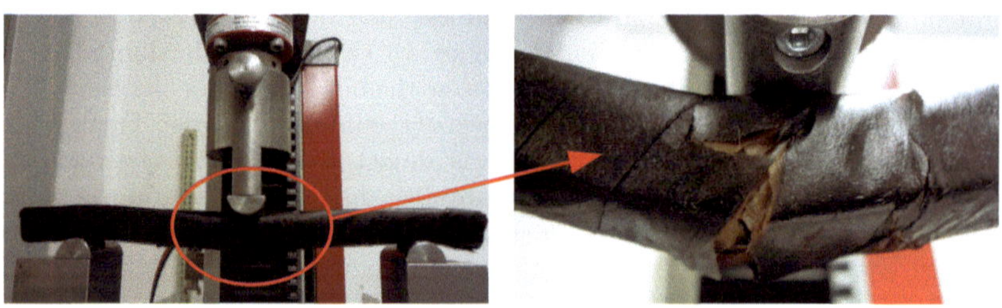

Figure 4.14: Bending test on a 90 days old CTC sample

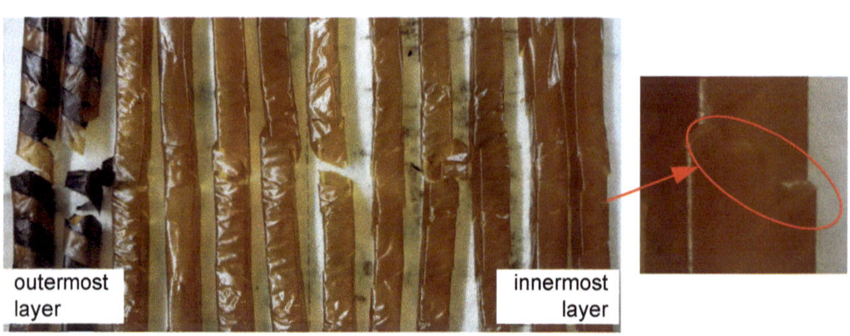

Figure 4.15: Paper strips from a 162 days old sample after bending test

5 N has been necessary, due to the lowered tensile strength. A shift of measurement results on the strain axis ensures that the virtual zero-crossings coincidence with the origin of the coordinate system, which improves the comparability of the curves. Figure 4.16 visualizes the wide-span stress-strain curves with error bars for the standard deviation, resulting from the measurement of 10 samples. The cross-sectional area of the paper strips is 30 mm × 85 µm.
All curves follow the same trend, which is typical for brittle materials: A quite linear increase in stress, followed by a maximum, and a fast decrease caused by the cracking of the samples. The ultimate tensile strain ϵ_u and the ultimate tensile strength σ_u characterize the maximum.

The curve for the factory new paper has two maxima at 2.38 % and 3.50 % strain. The first correlates to the beginning of cracking, but the paper is still stable enough to withstand a further force increase. After the second maximum, a large crack leads to the final rupture. However, the aged samples only show one maximum. With proceeding aging time both ϵ_u and σ_u decrease. For the outside layer this decrease is much more dramatic compared to the inner layers. Tables 4.2 and 4.3 list the results in mechanical strength for the inner and outside layers. The Young's modulus E_0 derives from the slope in the linear elastic range of the stress-strain curves. Not measurable entries are marked with n.a.

The degree of polymerization (DP) has been calculated from the zero-span tensile strength σ_u, based on a correlation developed in [84]. This method reduces the variance in DP estimation due to the large underlying measurement data set. The correlation between DP and normalized zero-span tensile strength is given in Appendix A.5. The DP calculation for the slightly increased σ_u (6 and 25 days) is not possible, as the underlying equation is only valid for $DP < 1300$.
The wide-span results unveil a significant paper aging because all results listed in Table 4.2 decrease with aging time. For the 162 days old samples the outer layers have almost lost their complete strength, as a reliable measurement has not been possible anymore. Compared to the inner layers the outer ones age must faster, which also is indicated by Figure 4.15.
The zero-span results from Table 4.3 allowed no distinct differentiation between inner and outer layers until the aging state of 90 days. Compared to the wide-span ultimate stress the zero-span results and the correlated aging rate are slightly lower. This leads to the conclusion, that the cellulose fiber network ages faster than the fibers itself. A significant decrease in DP starts from 90 days on. However, this parameter is not sensitive concerning the lost in mechanical (wide-span) strength.

4.4.4 Tensile and Bending Test Correlation

The following correlation is based on the critical deflection gathered from the bending tests according to Figure 4.13 and the ultimate tensile strength of the inner paper layers. Figure 4.17 a) visualizes the corresponding values with respect to the aging time. This graph shows, that the critical deflection better fits the trend

of the wide-span strength, compared to the zero-span results. Missing data points in critical deflection are the factory new and 6 days old results, as the underlying bending tests did not lead to measurable insulation damage.
Part b) of Figure 4.17 correlates the tensile strengths with the critical displacements. A linear curve fit of the wide-span data approximately runs through the origin, while the zero-span best-fit line has a significant axis intercept of 30 MPa. An ultimate tensile strength close to zero should result in almost zero critical deflection. This concludes, that the wide-span paper characterization is closer correlated to the bending strength of the conductor, compared to the zero-span method.

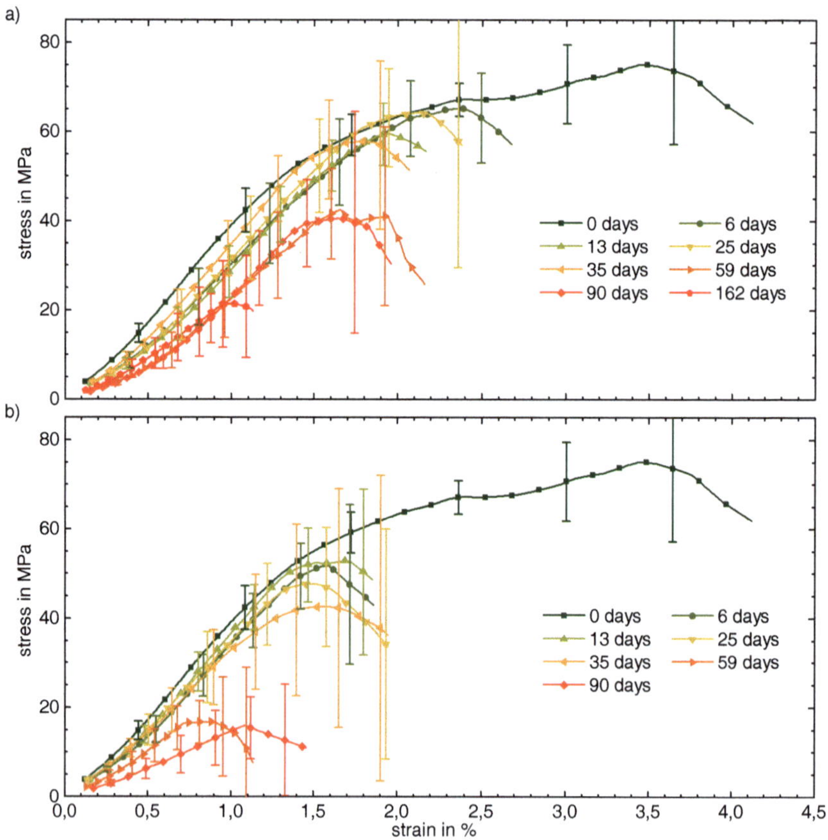

Figure 4.16: a) Wide-span measurement results for the 4 inner layers (MD)
b) Wide-span measurement results for the 4 outside layers (MD)

Table 4.2: Wide-span characterization of the aged insulating paper

days	ϵ_u in %		σ_u in MPa		E_0 in GPa	
	in	out	in	out	in	out
0	2.38		67.2		5.00	
6	2.38	1.57	65.1	51.8	4.20	4.36
13	1.92	1.46	59.5	52.0	4.43	4.76
25	2.11	1.45	64.2	47.7	4.32	4.29
35	1.78	1.52	57.8	42.6	4.75	3.80
59	1.66	0.74	42.5	16.8	3.75	2.86
90	1.59	1.08	40.5	11.6	4.07	1.81
162	0.95	n.a.	21.5	n.a.	3.13	n.a.

Table 4.3: Zero-span and DP characterization of the aged insulating paper

days	σ_u in MPa		σ_u in %		DP	
	in	out	in	out	in	out
0	54.1		100		1296	
6	58.5		108.1		≈ 1300	
13	49.5		91.5		745	
25	56.9		105.2		≈ 1300	
35	53.8		99.4		1233	
59	52.6		97.2		1047	
90	43.1	35.5	79.7	65.6	463	318
162	32.0	22.7	59.1	41.9	278	208

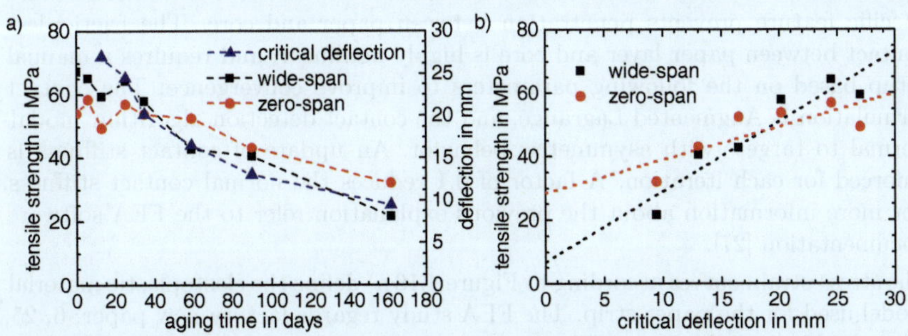

Figure 4.17: a) Critical deflection from bending tests and ultimate tensile strength (inner layers) over aging time. b) Correlation between critical deflection and ultimate tensile strength (inner layers)

4.4.5 FEA-Based Failure Analysis

The stiffness decrease deduced from the previous three-point bending tests enable a coarse estimation of the impact of cellulose aging. However, at the critical displacements marked in Figure 4.13, most of the paper layers already have been failed. In conclusion, it is not possible to evaluate the initiation of damage and its dependency from aging. The aim of the proposed FEA-based analysis is to correlate the strain arising in the layers as a result of conductor deformation to the wide-span tensile strain measured on the aged paper samples (Table 4.2).
The FEA model respects a paper covered conductor segment in the bending test arrangement, as visualized in Figure 4.14. The following simplifications reduce the FEA complexity, and make the problem convergent and solvable:

a) The strands are homogenized to a core.
b) The model considers one layer, which is wrapped without overlap. [91] provides additional FEA studies on straight CTC models covered by more paper layers.
c) Close to the supporting pins there is no paper, as high compressive stresses cause convergence problems. The paper ends are fixed to the core.
d) A displacement constraint in the middle of the core generates the conductor deflection, to overcome convergence problems caused by the contact surface between center stamp and paper layer.
e) The contacts between paper and core are frictionless.

Figure 4.18 shows the meshed FEA model for the CTC-B-21 three-point bending test beside the real test setup. Figure 4.19 visualizes the mentioned constraints and the bonding area between copper core and paper layer.
The discretization for the paper layer is based on shell elements. These kind of elements are suitable for thin to moderately thick shell structures [41, p. 377]. They offer the advantage that only one element in the thickness direction is necessary. A multizone meshing algorithm provides uniformly distributed elements. The discretization has to be small enough to resolve the rounded edges of the core. The inner core is meshed by solid elements. The shell-thickness effect as a FEA software specific feature prevents penetration between paper and core. The frictionless contact between paper layer and core is highly nonlinear, and requires a manual setup based on the following parameters to improve convergence: The contact formulation is Augmented Lagrange, and the contact detection algorithm "nodal-normal to target" with asymmetric behavior. An update of contact stiffness is enforced for each iteration. A factor of 0.1 reduces the normal contact stiffness. For more information about the previous explanation refer to the FEA software documentation [27].

The stress-strain curves according to Figure 4.16 a) define the elastoplastic material model used for the paper strip. The FEA study regards factory new paper, 6, 25, 59 and 162 days old paper. The core is a linear elastic isotropic material with a Young's modulus of 110 GPa. However, the material characterization for the homogenized core has been considered as irrelevant, as the displacement constraint in the conductor middle enforces the core deflection. The latter is ramped linearly

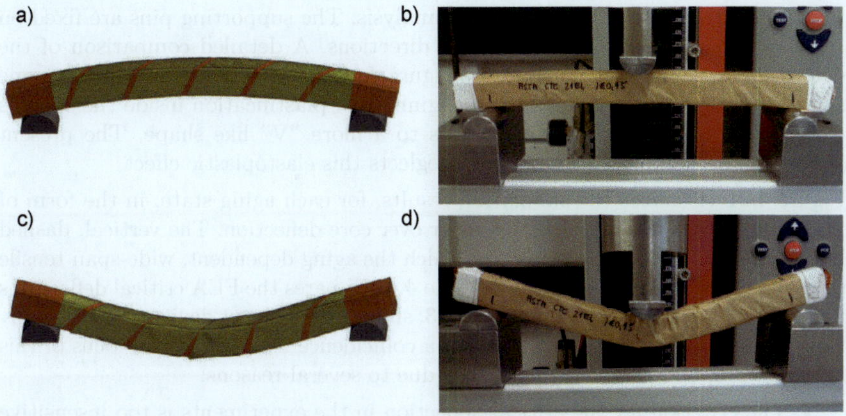

Figure 4.18: a) Undeformed, c) deformed bending test FEA model
b) Undeformed, d) deformed bending test sample

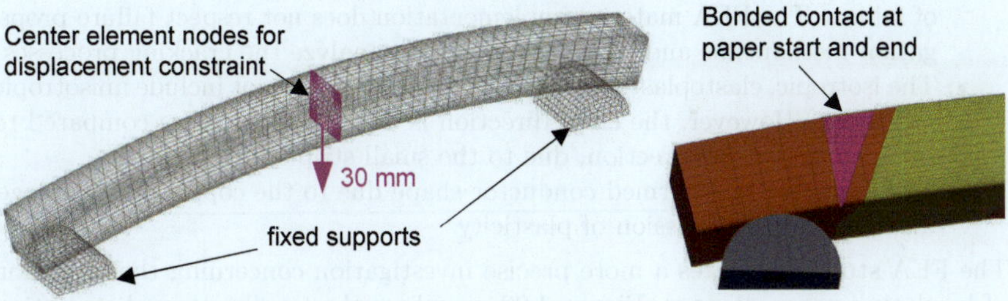

Figure 4.19: Constraints used for the FEA model

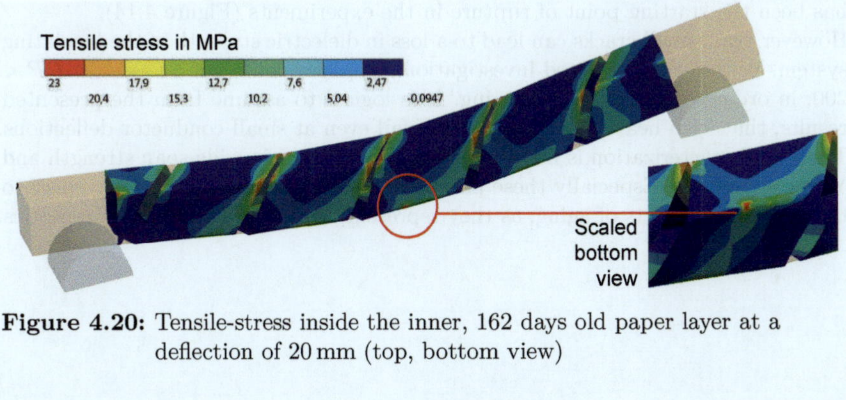

Figure 4.20: Tensile-stress inside the inner, 162 days old paper layer at a deflection of 20 mm (top, bottom view)

from 0 . . . 30 mm in a static structural analysis. The supporting pins are fixed on their lower surface in all three spatial directions. A detailed comparison of the deformed shape between simulation (Figure 4.18 c) and test (part d) shows some slight deviations. This is due to the beginning of plastification inside the strands close to the central stamp which leads to a more "V" like shape. The present simulation model for the copper core neglects this elastoplastic effect.

Figure 4.21 visualizes the simulation results, for each aging state, in the form of maximum tensile stress inside the paper over core deflection. The vertical, dashed lines mark the critical deflections, at which the aging dependent, wide-span tensile stress σ_u from Table 4.2 is reached. Table 4.4 compares the FEA critical deflections to the experimental results of Figure 4.13, showing significant deviations, especially for the slightly aged samples. However, a coincidence has not been in focus of this investigation, and is almost impossible due to several reasons:

- The criterion for the critical deflection in the experiments is too insensitive (the inner layers could not be observed).
- In the FEA study the layer is wrapped perfectly close to the core.
- There are no transposing strands.
- The critical deflection derived from the FEA results only mark the initiation of failure. The FEA material implementation does not respect failure propagation mechanisms and it is not possible to analyze the cracking processes.
- The isotropic, elastoplastic paper material model does not include anisotropic properties. However, the cross direction is difficult to measure compared to the manufacturing direction, due to the small strip width < 30 mm.
- The deviation in deformed conductor shape due to the copper core homogenization and the omission of plasticity.

The FEA study facilitates a more precise investigation concerning the initiation of insulating paper damage. Figure 4.20 visualizes the tensile stress distribution inside the 162 days old paper for a deflection of 20 mm. Even for this deflection, significantly exceeding the experimental critical deflection, σ_u is only reached in a small area marked by the red circle. This point close to the conductor center also has been the starting point of rupture in the experiments (Figure 4.14).

However, even small cracks can lead to a loss in dielectric strength of the insulating system. Within the presented investigation the paper could not be aged to $DP <$ 200, in order to address severe aging. It is logical to assume from the presented results, that such heavily aged paper will fail even at small conductor deflections. The DP characterization is not able to identify the loss in wide-span strength and Young's modulus. Especially these parameters are very important with respect to the mechanical effects of aging, as they represent important mechanical properties.

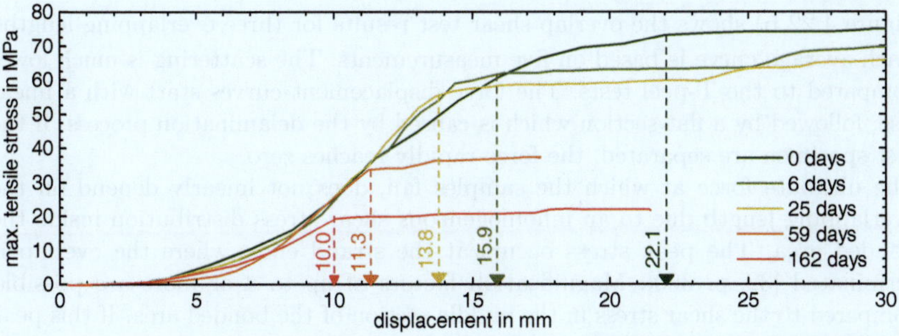

Figure 4.21: Max. tensile-stress inside the innermost paper layer over stamp displacement derived from the three-point bending FEA model

Table 4.4: Critical deflection from experimental and FEA-based results

	factory new	6 days	25 days	59 days	162 days
Test	> 30 mm	> 30 mm	24.5 mm	16.5 mm	9.5 mm
FEA	22.1 mm	15.9 mm	13.8 mm	11.3 mm	10 mm

4.5 Characterization of Epoxy Bonded CTCs

This section analyzes the thermosetting epoxy coating used for bonded CTCs by the means of T-peel and overlap shear tests, as described in Section 4.1. The developed FEA model allows investigating the behavior of bonded CTCs under radial buckling forces. The presented measurements are based on preliminary work published in the author's publication [90].

4.5.1 Test Results

The proposed measurement results are based on epoxy coated strands from CTC-A-11, according to Table B.2. Figure 4.22 a) visualizes the T-peel test results deduced from five measurements. The displacement axis is scaled logarithmically to enhance clearness. The peak of the curves shows significant scattering between 65 . . . 100 N. The delamination starts instantaneously after this peak force. The variation in peak force implies that a reproducible test specimen fabrication or the uniformity of the adhesive coating are not perfect. However, it has not been possible to reach further improvements to this point. The subsequent force decay after the peak has a significantly lower variation between the measurements. Within the correlated displacement range, the delaminated length increases.

Figure 4.22 b) shows the overlap shear test results for three overlapping lengths. Each average curve is based on five measurements. The scattering is much lower compared to the T-peel tests. The force-displacement curves start with a linear rise, followed by a flat section which is caused by the delamination process. If the test specimen are separated, the force rapidly reaches zero.
The ultimate force at which the samples fail, does not linearly depend on the overlapping length due to an inhomogeneous shear stress distribution inside the bonded area. The peak stress occurs at the strand ends, where the overlap is terminated [32, p. 300]: Magnification factors of up to 4 or more are possible, compared to the shear stress in the middle section of the bonded area. If this peak stress exceeds the shear strength of the adhesive, the delamination starts.
This indicates that the calculation of a maximum shear strength from ultimate force F_u and overlap area A_o according to $\tau_{max} = F_u/A_o$, leads to values, which are strongly dependent on the overlap length, as shown in Table 4.5. For calculating A_o, the strand width h_s has been reduced by twice the edge radius of 0.4 mm, which reduces the bonding surface.

Beside these measurements at room temperature, additional shear tests have been carried out at a worst-case temperature of 120 °C. A heating tube around the test specimens, and a thermal sensor applied on the overlapping area were used to control the temperature. The resulting ultimate force given in Table 4.5 proves that there is no dramatic difference referring to the room temperature measurements. Only in the case of 6 mm overlap, a decrease in shearing strength by 30 % occurs. As the overlapping lengths within CTCs are much higher, this is ignored. Furthermore, the manufacturer allows a reduction by 50 % in shearing strength for a rise from room temperature to 125 °C. The subsequent FEA based evaluation is done for the 20 °C data, as the copper stress-strain characteristic also has been measured at room temperature. Due to the thin strands as carrier object for the adhesive coating, the copper elastoplastic properties contribute to the measurement results.

Table 4.5: Shear strength results deduced from the overlap shear tests

Overlap in mm	Temperature in °C	F_u in N	τ_{max} in MPa	Std.dev. in MPa
6	20	678.37	22.84	1.27
6	120	465.08	16.06	1.37
12	20	976.27	16.99	1.50
12	120	987.75	17.54	1.20
24	20	1208.09	10.22	0.62
24	120	1130.32	9.22	0.37

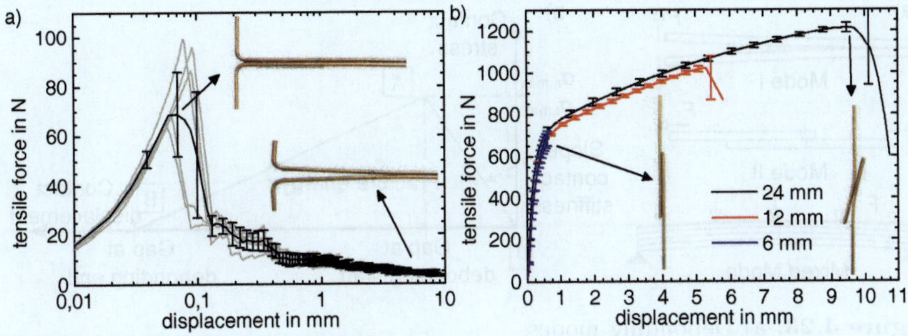

Figure 4.22: a) T-peel-test results
b) Overlap shear-test results for different overlap lengths

4.5.2 Adhesive Layer Parameterization

The adhesive layer is modeled based on a cohesive zone model (CZM). This approach is one of the most widely used tools to simulate adhesives and their failure with the help of FEA software. The CZM approach is suitable wherever the strength of an adhesive layer is relatively weak compared to the adjoining material [46]. This section only provides a brief description of the important parameters. For more details refer to the referenced literature.

The mesh elements associated with the CZM describe the constitutive relation between the stress arising in the interface and the corresponding interfacial separation. Figure 4.23 a) visualizes the debonding mechanisms: Contact separation in normal direction (Mode I), in tangential direction (Mode II) and a combination of both (Mixed Mode).

One of the most commonly used underlying constitutive equations is the bilinear law, as proposed in [44]. Figure 4.23 b) provides the corresponding stress-displacement jump relation. This relationship is related to both tangential and normal direction. The line (0A) describes the linear elastic loading of the contact interface, whereas its slope describes the contact stiffness in normal or tangential direction K_n, K_t. The maximum stress $\sigma_{\mathrm{n,max}}$ or $\sigma_{\mathrm{t,max}}$ is reached in A, where the contact debonding starts. During the subsequent linear softening (AB) the contact stiffness is progressively reduced to zero. The slope of (AB) characterizes whether the fracture is ductile or brittle. Point B characterizes the debonding end, as the contact stress reaches zero. The area under the traction displacement jump relation is equal to the fracture toughness (or critical fracture energy release rate) in normal and tangential direction G_n, G_t.

The radial bending of CTCs is a mixed mode problem with dominating tangential debonding. In conclusion, the characterization of both normal and tangential CZM parameters is evident, but the focus is on the tangential parameters.

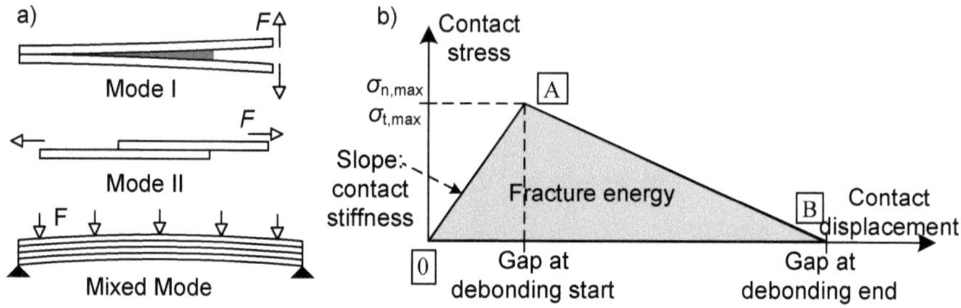

Figure 4.23: a) Debonding modes
b) Constitutive relationship for the bilinear CZM approach

The adherent copper and the adhesive coating are isotropic materials. Therefore, it is possible to equalize the normal and tangential stiffness $K_{\mathrm{n}} = K_{\mathrm{t}} = K$. The following set of five parameters describes the stress-displacement jump relation in normal and tangential direction: $K, G_{\mathrm{n}}, G_{\mathrm{t}}, \sigma_{\mathrm{n,max}}$ and $\sigma_{\mathrm{t,max}}$.
In literature, there exist no clear guideline or approach to estimate CZM parameters [63]. Especially concerning the stiffness K of the cohesive elements, there is no general agreement. The authors of [63] propose to approximate the stiffness from the Young's modulus of the adhesive layer and its thickness. Other authors propose values between $10^4 \dots 10^7\,\mathrm{N\,mm^{-3}}$ [50]. It is accepted, that variations in this parameter do not affect the final results too much. A decrease in the cohesive zone stiffness improves the solution convergence.

The following CZM parameter estimation is based on the method of inverse FEA: The simulation model results regarding the T-peel and overlap shear test setups are adapted to the measured force-displacement characteristics. The aim is to reproduce all experiments with one consistent set of CZM parameters.
Figure 4.24 shows some exemplary FEA study results. The geometry is based on the CTC-A-11 strand dimensions including the edge radii of 0.4 mm. Elastoplastic material behavior is assumed for the copper strands, as their stiffness influence the debonding characteristic. A rigid constraint fixes one end of the test specimen. A displacement constraint applied to the other end reproduces the travel of the testing apparatus clamping jaw. For a reliable comparison of force-displacement characteristics between experiments and FEA study it is important to respect the full clamping length in the FEA, which is not part of the cut out views of Figure 4.24.
A useful compromise between convergence speed and reproduction of measurement data has been found for a stiffness $K = 5000\,\mathrm{N\,mm^{-3}}$. The remaining CZM normal parameters have been estimated from the T-peel test results. The tangential values have been derived from the shear tests. However, there is a mutual influence: The shear test also introduces normal stress components in the adhesive layer due to the incline of delaminated strand ends, as Figure 4.24 b) shows.

Figure 4.25 visualizes some exemplary parameter variations and their effects for the T-peel test. The maximum normal stress $\sigma_{n,max}$ mainly has an impact on the peak force, as shown in part a). A change in normal fracture energy, however, both influences peak force and subsequent delamination behavior, as part b) of Figure 4.25 states. The parameter set $\sigma_{n,max} = 15\,\mathrm{MPa}$, $G_n = 0.2\,\mathrm{kJ\,m^{-2}}$ reproduces the average measurement curve satisfyingly.
However, the FEA studies on the shear test model figured out that a normal fracture energy below $1\,\mathrm{kJ\,m^{-2}}$ leads to serious convergence problems. Therefore, $G_n = 1\,\mathrm{kJ\,m^{-2}}$ has been used, and the deviations in the normal delamination behavior have been accepted regarding the focus on a precise tangential delamination behavior. Figure 4.26 shows the best match between FEA studies and experimental results for the shear tests, based on the final CZM parameter set according to Table 4.6.

Table 4.6: CZM parameters

$\sigma_{n,max}$	$\sigma_{t,max}$	G_n	G_t	K
10 MPa	30 MPa	$1\,\mathrm{kJ\,m^{-2}}$	$2.5\,\mathrm{kJ\,m^{-2}}$	$5000\,\mathrm{N\,mm^{-3}}$

4.5.3 FEA Model Validation

Three-point bending tests on straight, bonded CTC samples have been carried out, to verify the presented CZM approach. According to the CTC manufacturer guideline, thermal activation of the adhesive has been done in an oven at 120 °C for 24 h, under a constant pressure of 1 MPa. Each five samples of CTC-A-11 have been tested at a support distance of 100 mm and 180 mm.
The FEA geometry contains the two supporting stamps, which are constraint by a fixed support on their lower face. Figure 4.27 shows the meshed 100 mm model and an exemplary deformation. A displacement constraint moves the test stamp centrally towards the CTC. The latter is included by the means of the simplified model with straight strands without transpositions. The strand edge radii are considered. CZM contacts between the strands characterize the bonding, based on the parameters according to Table 4.6. Contact behavior is symmetric which means that the contact surfaces are constrained from penetrating the target surfaces and the target surfaces are constrained from penetrating the contact surfaces. The contact formulating algorithm is pure penalty. Tangential slip under normal compression is activated, to ensure proper mixed mode contact debonding. The contacts between CTC and stamps are frictionless. The stamp material is steel, while an elastoplastic stress-strain curve characterizes the strands.

Figure 4.28 provides the FEA results as stamp reaction force in the direction of travel over the applied displacement (dashed lines). The solid curves correspond to the averaged results from the three-point bending tests. For both support distances,

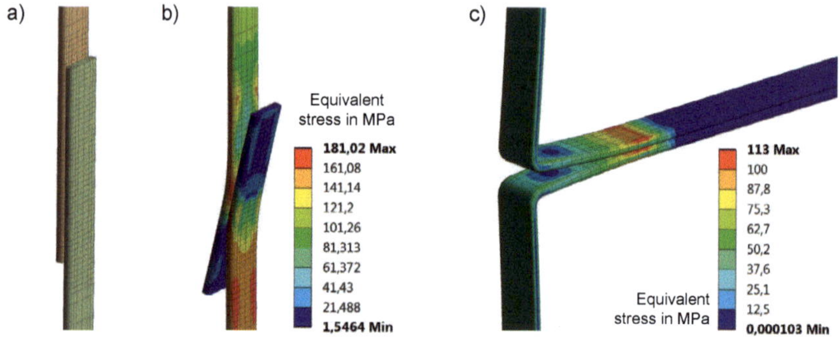

Figure 4.24: a) Shear test 24 mm overlapping FEA model (cut out view)
b) Shear test FEA model shortly before complete debonding
c) T-peel test FEA model at a delamination of 20 mm

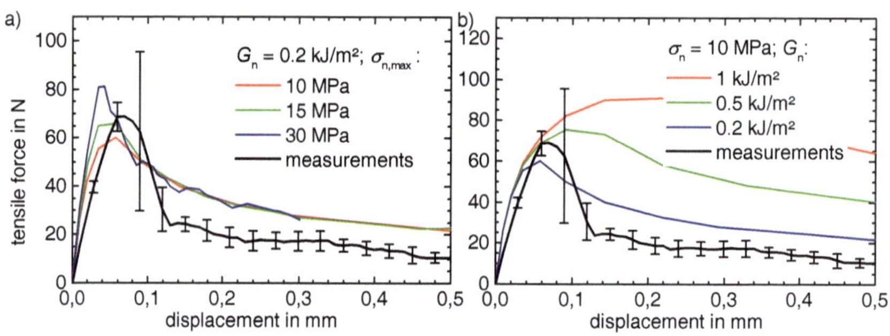

Figure 4.25: T-peel test vs. FEA results for constant tangential CZM parameters according to Table 4.6 and variation in normal parameters: a) Variation in $\sigma_{n,max}$, b) variation in G_n

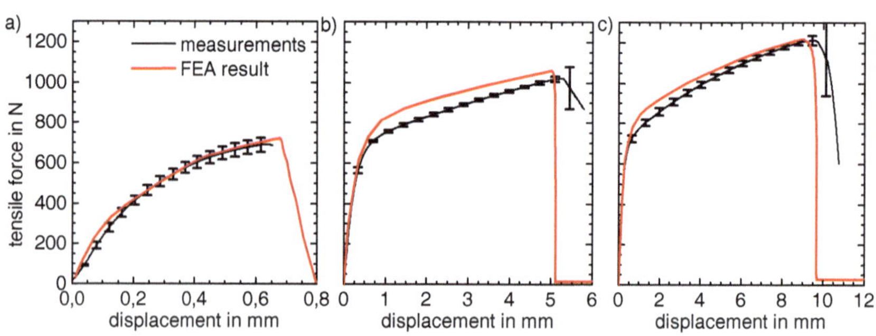

Figure 4.26: Shear test vs. FEA results with CZM parameters according to Table 4.6 for an overlap of a) 6 mm, b) 12 mm and c) 24 mm

a convenient coincidence is reached. Only for displacements above 10 mm, the reaction force deduced from the FEA studies is slightly below the experimental data. The deviations at high deflections arise from frictional forces between conductor and stamps which are not included in the FEA models. Furthermore, the simplified conductor model contributes to the deviations.
Regarding the equivalent von Mises stress arising in the adhesive layers (red dotted lines), both cases reach the critical shear stress of $\sigma_{\mathrm{t,max}} = 30\,\mathrm{MPa}$. This concludes that some strands start to delaminate. Figure 4.29 visualizes the equivalent stress in an exemplary CZM layer at the end deflection for the 180 mm case. The peak stress mainly occurs close to the middle stamp and the outer CTC boundaries. This indicates that only a few small spots are affected by delamination. A visual inspection of the bent test specimen also did not unveil a complete delamination.

4.5.4 FEA Model for Bonded CTCs

Finally, the buckling behavior of bonded CTCs is investigated based on a detailed CTC model including transpositions, which already has been used in Section 3.2.1. The underlying geometry represents one sector from a transformer winding between two axial sticks. The latter are not included in the simulation, but replaced by a clamped constraint at the sector endings.
The CTC strand dimensions are from CTC-B-21 according to Table B.2. Two different number of strands are considered: $N_{\mathrm{s}} \in \{11, 21\}$. The removal of edge rounding reduces the model complexity. The transposition pitch is 60 mm. Table 4.7 lists the remaining geometric parameters and the model labels. The dimensionless rise h_{r}^* refers to the total CTC thickness in the radial direction, and not only to the strand thickness t_{s}.
The material is of elastoplastic co-65 according to Table B.1. Cohesive elements with CZM parameters from Table 4.6 define the contact surface between the strands. The remaining contact configuration is equal to the previous three-point bending FEA study. The radially directed, magnetic force is applied in the form of a pressure on each strand, acting in the radial direction.

Figure 4.30 visualizes an exemplary simulation result for the bonded CTC model f with 11 strands. The left part corresponds to a load before the initiation of buckling. It is obvious, that the bonding is still intact, as it transfers stresses between the strands without any discontinuities. The right model shows the state after buckling. Discontinuities in the stress distribution between the strands unveil a partial delamination of the adhesive layers.
Figure 4.31 plots the conductor deflection and the maximum contact stress in the CZM elements for the four variants of the 11 stranded CTC. The vertical solid lines represent the analytical critical loads, calculated by Equations (3.20) and (3.28) with elastoplastic material properties as described in Section 3.1.5. The analytic approach implies a perfect bonding which does not fail.

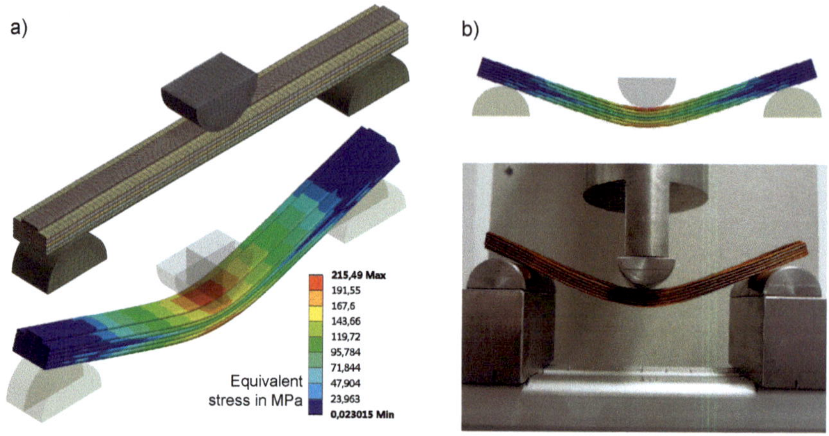

Figure 4.27: a) 3D FEA model of 100 mm bending test
b) FEA model and test at a stamp displacement of 15 mm

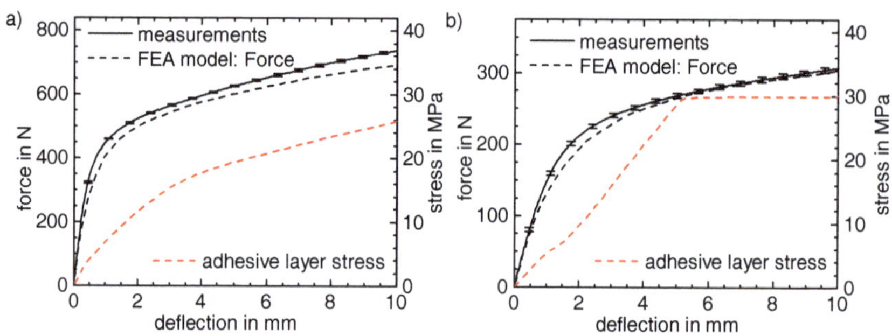

Figure 4.28: Comparison between measurement and FEA results
a) 100 mm bending test, b) 180 mm bending test

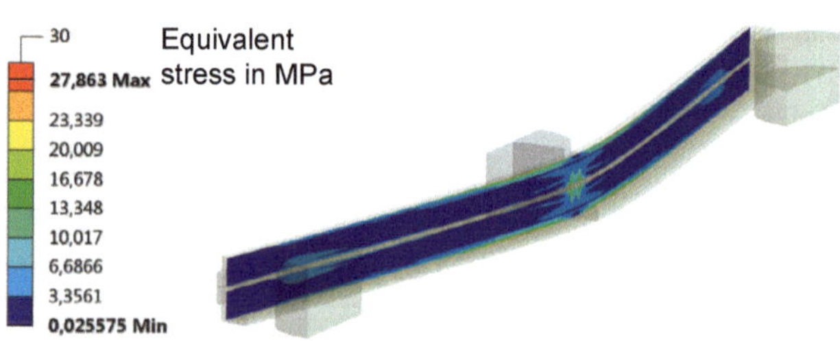

Figure 4.29: Equivalent stress arising between to pairs of strands in the 180 mm test, at a deflection of 15 mm

The variant f with 18 spacers is the only one showing distinct buckling. However, the normalized rise h_r^* is smaller than 2.85, which is, according to Section 3.1.4, an indicator for pure bending. This deviation results from the assumption of linear elastic material for the normalized rise criterion. The critical load from the FEA model is lower than the analytical load, due to the initiation of delamination. The maximum stress inside the adhesive layers proves this assumption, as it reaches the limit of 30 MPa beyond the theoretical load.
With an increasing number of spacers the sharp transition between original and buckled state changes to a smooth deflection. Determining a critical load is not clear, and the analytic loads have no relevance. For these cases the normalized rise h_r^* is significantly smaller than 2.85, indicating the pure bending failure process. The stress inside the bondings only reaches the delamination limit for large deflections.
In the case of 21 strands (Figure 4.32) the models d, e and f show distinct buckling, although $h_r^* < 2.85$. The critical loads are greatly exceeding the analytic values. Model c even did not fail for the considered load amplitude. However, this time, the critical stress in the adhesive layers is reached for a load smaller than the critical load. In conclusion, the beginning of delamination is responsible for buckling initiation.
For the buckling strength of thermosetting CTCs, it is evident to know the critical load which initiates delamination of strands. The proposed FEA models provide a precise estimation method to this kind of problem. The analytic elastoplastic buckling loads are too low if the conductor fails in pure bending or too high if buckling occurs which then is affected by delamination. To conclude, the application of analytic buckling formulations like Equation (3.19) based on a bending stiffness according to Equation (3.5) is erroneous.

Table 4.7: Geometry parameters for bonded CTC models

model label	R in mm	N_{as}	α in °	h_r^* (11 strands)	h_r^* (21 strands)
c	515.6	40	9	0.41	0.22
d	515.6	30	12	0.73	0.38
e	515.6	24	15	1.15	0.60
f	515.6	18	20	2.04	1.07

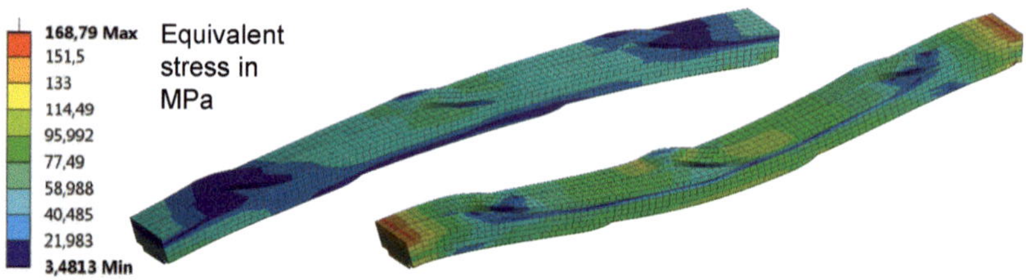

Figure 4.30: a) Equivalent stress before and after buckling (Model f, 11 strands)

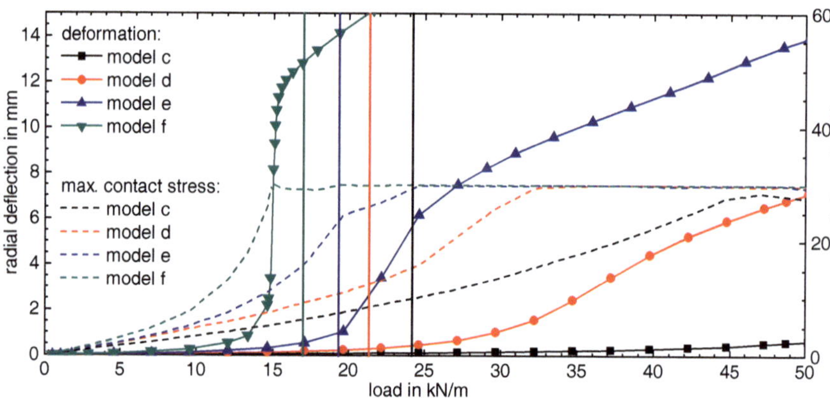

Figure 4.31: Simulation results for CTC-A-11

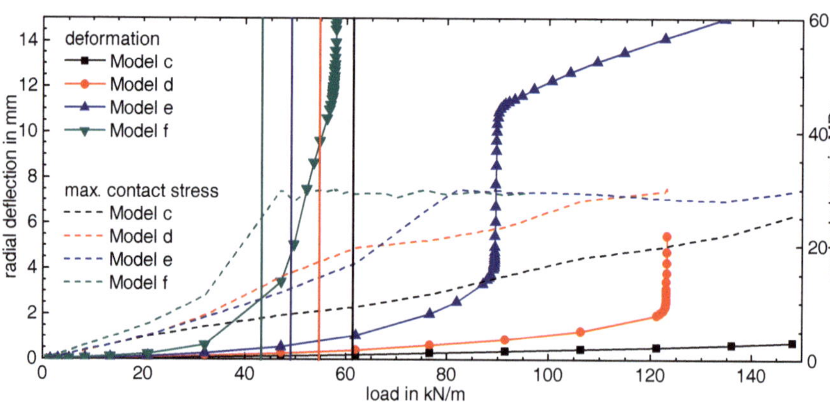

Figure 4.32: Simulation results for CTC-A-21

Chapter 5

Dynamic Short-Circuit Forces Test Stand

5.1 Design and Principal Functionality

The purpose of the test stand, proposed in this chapter, is to evaluate the buckling strength of power transformer inner windings against radial forces. The setup only regards a part of an inner winding, as it is not practical to do buckling experiments on a complete transformer coil. Figure 5.1 shows the essentials of the test stand with inner and outer winding wound around an iron core. Both windings are connected in series, to keep the ampere-turns balanced. Therefore, the current has to flow in opposite direction between both coils. This direction is equal to the real transformer, as shown in Figure 2.6. The core is made from a hollow cylinder, to cut down weight and material consumption. It has two functions: It supports the inner windings, and guides some of the leakage flux close to the inner side of the inner coil. The latter increases the radial force arising on the inner winding. A radial sheeted construction, made from iron sheets, reduces eddy currents.
Figure 5.2 shows a cut-through view: Axial sticks support the inner winding against the core. Axial sticks between both coils support the outer against the inner winding. The radial spacers separate the turns in the axial direction. Threaded bolts provide a clamping force in the axial direction. As these parts are made from iron, it is necessary to evaluate their disturbing effect on the stray flux distribution between the coils. FEA-based studies proved that this effect is negligible.

The clamping supports shown in Figure 5.3 a), b) provide the electrical connection and mechanical fixation of the conductors. To homogenize the contact resistance between clamping support and CTC, the strands insulating coating may only be removed on the top of the upper stack, and on the bottom of the lower stack. The remaining strand faces remain insulated. Otherwise, a much higher current will flow through the outer strands compared to the inner strands. The tightening torque

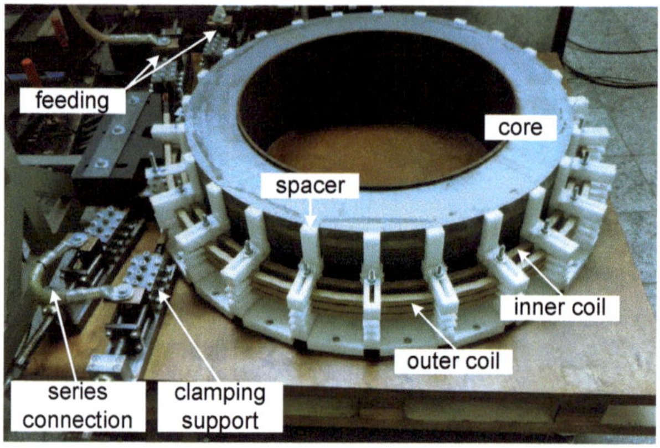

Figure 5.1: Electromagnetic short-circuit forces test stand

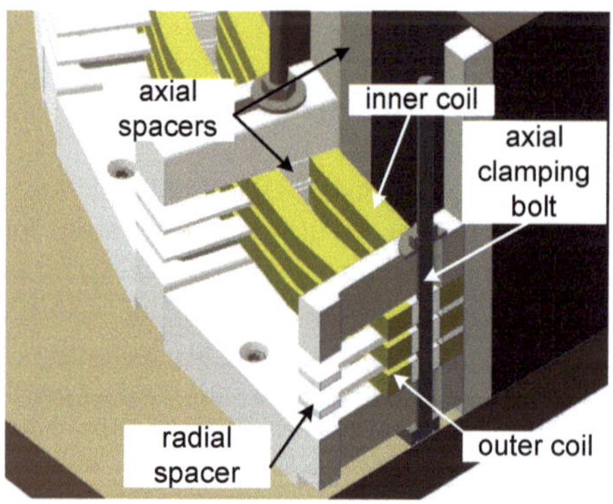

Figure 5.2: Cut through CAD view of the test-setup

of the greased trapezoid thread adjusts the winding pretension. The correlation between tensile force F and tightening torque M is $F/M = 450\,\mathrm{m}^{-1}$.

A three-phase high-current power transformer, which can deliver up to 40 kA RMS at a maximum phase-phase voltage of 245 V RMS, supplies the test stand. The supply voltage is variable in twelve steps between 74 ... 245 V. A variable high-current inductor of 1.47 ... 16.8 µH connected in series between transformer and test stand provides a further adjustment of test current.
A combination of a fast thyristor-based switch and magnetic-driven high-current switch triggers the connection between supply transformer and test stand. For a detailed description refer to [95]. The thyristor switch allows a precise, phase-correlated connection. The parallel connected mechanical switch closes a few milliseconds later, and due to its lower impedance the current commutates into this

switch which has a better long term current carrying capacity, but a significantly varying delay time. The phase angle of the feeding voltage at the switch-on-point influences the first peak current amplitude, as explained in Section 2.2.1.

A typical test duration is 200 ... 300 ms to reduce a heating up of the coils. First results showed that maximum winding deformation is directly correlated to the maximum force, which itself is related to the peak current occurring in the first half wave of the feeding voltage. Testing starts with a low current to stay within elastic winding vibrations. In subsequent tests, the test current is sequentially increased until significant buckling occurs. A high-speed camera windings measures conductor deformations in the radial direction. Furthermore, an indirect measurement of vibration amplitudes using acceleration sensors has been implemented. A Rogowski sensor records the test current $i_k(t)$. FEA simulations regarding the magnetic flux distribution estimate the electromagnetic forces from $i_k(t)$. Figure 5.4 visualizes a block diagram of the test stand functionality and measurement data processing steps. For more details concerning the test stand design refer to [87].

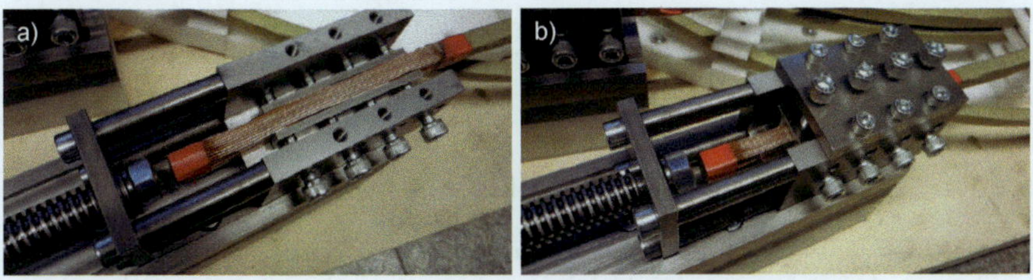

Figure 5.3: a) Opened, b) closed clamping support for the winding endings

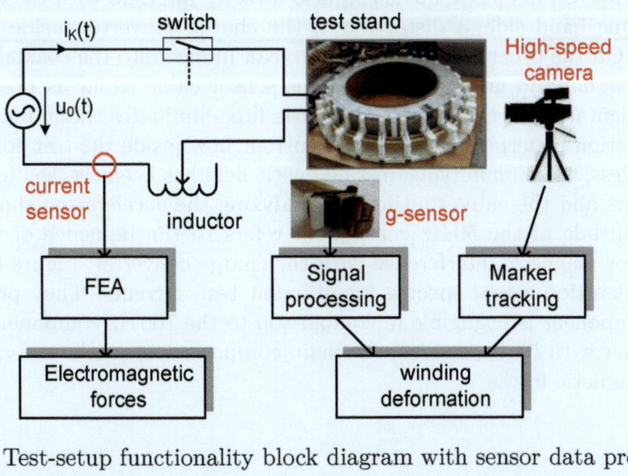

Figure 5.4: Test-setup functionality block diagram with sensor data processing

5.2 Deformation Measurement Systems

A measurement of winding deformation while the test current flows is only possible with a system which is immune against high magnetic fields. This concludes that electrical strain gauges are not applicable. Although there exist fiber optical systems, another problem is the undefined mounting on the insulating paper. Especially for CTCs with many layers of insulating paper, the strain induced into the outer layer is highly random and unreliable. As a solution, an indirect measuring method based on acceleration sensors, and a direct but contactless measuring method based on a high-speed camera have been developed. Student thesis [96] describes the preliminary investigations regarding the acceleration sensor system.

5.2.1 Acceleration Sensors

Shear-type piezoelectric acceleration sensors work after the principle of charge-separation inside a piezoelectric crystal, which deforms under the influence of sensor movements. The detailed construction and functionality is explained in [13, 96]. The data sheets Figures B.1 and B.2 provide the technical and calibration data of the used sensors 8772A50 (50 g) and 8728A500 (500 g) from Kistler Instrumente AG.

5.2.1.1 Immunity to Magnetic Fields

Piezoelectric sensors are most widely immune to magnetic fields, if the crystal inside is made from a ceramic which has no ferromagnetic properties. Otherwise, a deformation of the ceramic due to magnetostriction will distort the sensor output signal. However, magnetic fields can also directly interfere with the output signal: On the one hand side, a distortion of the charge converter inside the sensor is possible. On the other side, loop currents can induce into the coaxial sensor cable between signal and ground. Routing the sensor cable as far as possible parallel to the orientation of the leakage magnetic flux eliminates the induction problem. This direction is perpendicular to the current flow inside the test coils.
Nevertheless, the immunity against magnetic field has been checked for the individual sensors and the cable routing, by analyzing the acceleration signal spectrum. The amplitude of the 50 Hz component refers to the influence of induced loop currents or magnetic interference with the charge converter. Figure 5.5 visualizes four acceleration signal spectra for different test currents. They prove that the 50 Hz component is negligible in comparison to the 100 Hz component. The latter is, together with its harmonics, the main component of conductor vibrations due to the magnetic forces.

5.2.1.2 Mounting to the Windings

Mounting acceleration sensors to a test object is very crucial regarding the resonance frequency of the overall measurement arrangement [78]. A rigid mounting, for example by a threaded bolt, increases the resonance frequency in contrast to a weak mounting with, for example, double-sided tape. Therefore, it is not recommended to glue the sensor onto the outer insulating layer of the conductor. The coupling between strands and paper layers is very weak and undefined.
The approach used in this work is to press the sensor against the CTC by the means of a cable tie and a mounting cap (Figure 5.6). This solution compresses the paper layers between sensor and copper strands, and improves the coupling. Figure 5.7 clearly proves the difference: The red measurement data results from a glued sensor, the black data correlates to the cable tie variant. For most of the frequencies above 100 Hz the spectrum measured with the glued sensor is significantly lower compared to the cable tie variant. Also the important 100 Hz component is influenced.

5.2.1.3 Signal Processing

It is a well-known fact, that the integration of an acceleration signal leads to a velocity signal, which itself results after a further integration in a displacement signal. However, for the application with acceleration sensors, this method will lead to enormous errors. Small offsets or time-invariant drifts of the sensor's zero signal will cause a linearly with time increasing error in velocity, and a quadratic with time increasing error in displacement [23].
An example of such a faulty integration is given with [48]. The authors used acceleration sensors to measure the displacement of a transformer coil during short-circuit, without additional signal processing. Their simple double-integration process estimated deformation signals which increased quadratic with time.
To overcome this problem, three signal processing methods to remove constant and low-frequency drifts before each integration step have been developed and compared against each other. Figure 5.8 shows the corresponding block diagrams.

- The high pass filtering method a) adds a zero-phase Butterworth filter of fourth order with a cut-off frequency of 10 Hz in front of each integrator.
- In method b) an average and linear detrend filter removes constant offsets and linear trends before the integrators.
- Method c) is a filtering algorithm which works in the frequency domain, adapted from [28]. This approach overcomes aliasing effects in the time domain, which occur if low-frequency spectral components are set to zero.

The following steps and equations based on a generic time signal $w(t)$ explain method c) in more detail. The basis is the signal spectrum W, calculated by a Fast-Fourier-Transformation (FFT):

$$W = \mathrm{FFT}(w) \tag{5.1}$$

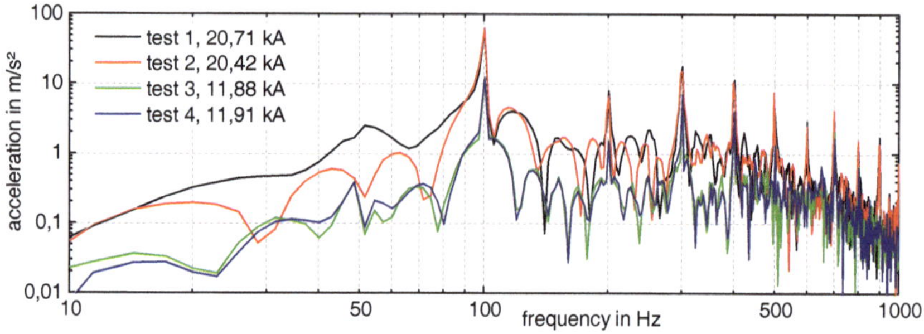

Figure 5.5: Acceleration signal spectra from the 500 g sensor mounted to the inner middle winding from type CTC-B-21

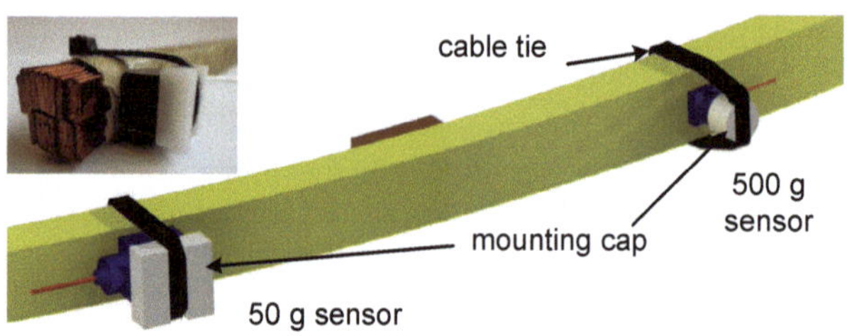

Figure 5.6: Rendered view of the mounting method for the acceleration sensors

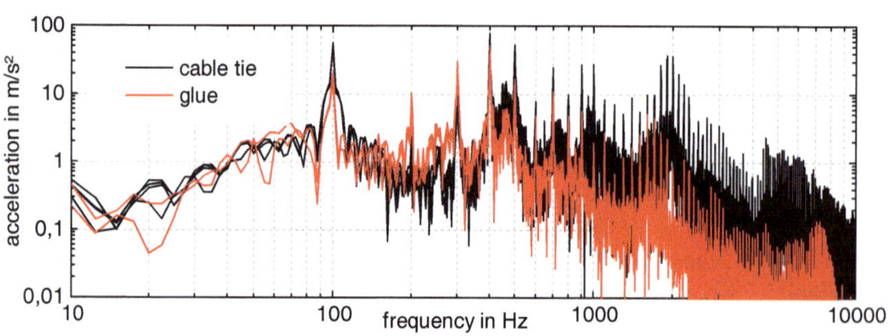

Figure 5.7: Acceleration signal spectra for different sensor mounting methods using a paper insulated CTC at 10 kA peak (elastic vibration)

The first component $W(1)$, which represents the constant component, and components $W(2)\dots W(n_c-1)$ are replaced by the n_c-th component $W(n_c)$, as shown in Equation (5.2). The cut-off index n_c corresponds to a frequency which corresponds to $n_c \cdot \Delta f$ with FFT frequency resolution $\Delta f = [\text{sample time}]^{-1}$.

$$W(k) = W(n_c) \quad \text{for } k = 1\dots n_c - 1 \tag{5.2}$$

Equation (5.3) restores the symmetry of the complex spectrum. The index N corresponds to the dimension of complex spectrum.

$$W(N-k) = \text{conj}(W(n_c)) \quad \text{for } k = 0\dots n_c - 3 \tag{5.3}$$

Finally, the Inverse Fast-Fourier-Transformation (IFFT) provides the filtered time signal w_f.

$$w_f = \text{IFFT}(W) \tag{5.4}$$

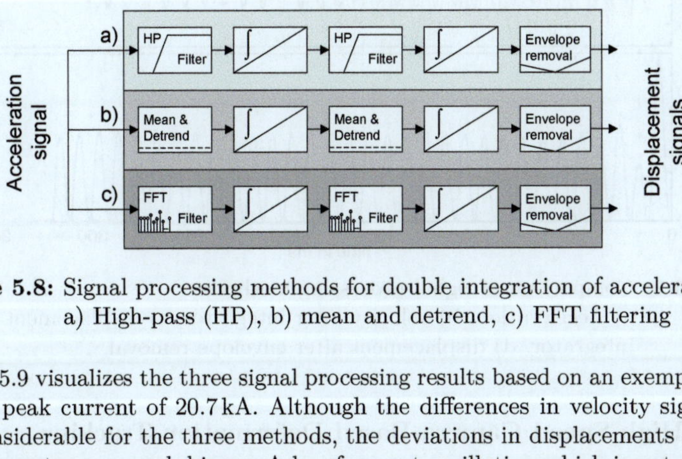

Figure 5.8: Signal processing methods for double integration of acceleration: a) High-pass (HP), b) mean and detrend, c) FFT filtering

Figure 5.9 visualizes the three signal processing results based on an exemplary test with a peak current of 20.7 kA. Although the differences in velocity signals are not considerable for the three methods, the deviations in displacements after the last integrator are much bigger: A low-frequent oscillation which is not plausible superimposes the signals provided by high pass method a) and detrend method b). The FFT filtering signal path c) leads to the most reliable displacement results. However, the residual deformation at the test end shows an unreasonable rising trend for the FFT method.

A finalizing lower envelope removal applied to the displacement signal improves the comparability of the three methods. The lower envelope calculates from the displacement signal minima and has to be subtracted from the displacement signal. Part d) shows the corresponding results. The deviation between all signal processing steps is close to zero. Section 5.2.3 provides a further analysis of the topic of residual deformation at the test end by a comparison to High-Speed camera measurement results.

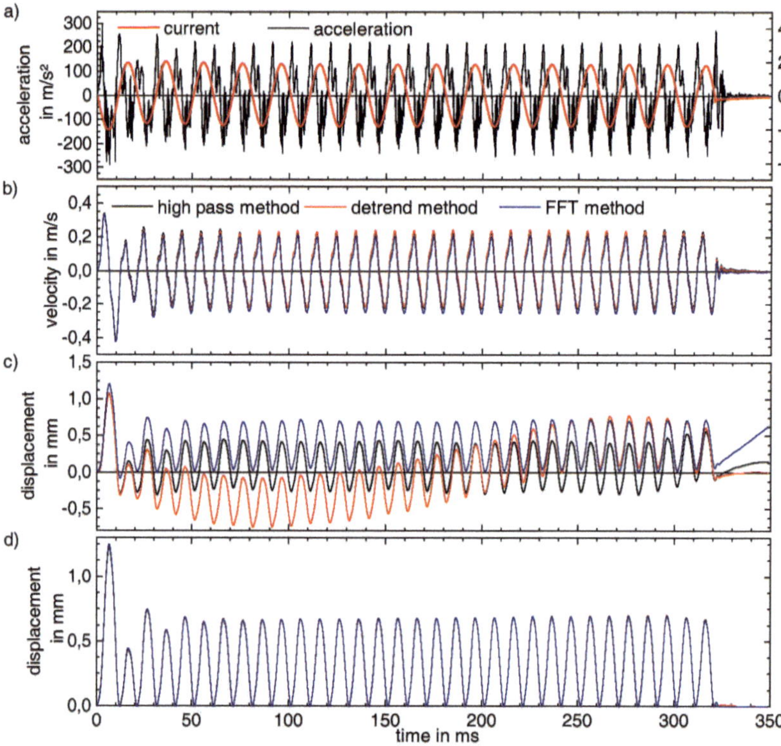

Figure 5.9: Comparison of signal processing methods:
a) Acceleration, b) velocity after integrator, c) displacement after integrator, d) displacement after envelope removal

5.2.2 High-Speed Camera-Based Deformation Tracking

5.2.2.1 Principal Functionality

The acceleration sensor measurement system is well suitable for small elastic winding vibrations. However, it is limited concerning large residual deformations at the test end. In conclusion, a contact-free optical measurement system has been developed. The high costs of laser vibrometers, and their limitation to only one measurement point confirmed the focus on a high-speed camera-based system. The 100 Hz-periodic vibrations require frame rates above 1000 Hz for a smooth sampling of deformations. Preliminary work to this topic is given with [89].
The high-speed camera records a coil sector between two axial sticks from above, as shown in Figure 5.10. Motion tracking markers are placed on the interesting

conductor positions. As the radial winding vibration has the highest amplitude in the sector middle, a tracking marker is placed to this position. For measuring vibrations at middle and lower turns, which are covered by the overlying conductor, the markers are fixed on mounting plates. A reference marker fixed to the iron core compensates for residual movements from low-frequent camera or test stand fluctuations. A scale near to the markers allows calibrating the camera recordings. Placing the scale in the same plane near to the markers avoids perspective distortions. A commercial motion tracking software estimates the marker displacements from the camera recording. However, the presented camera arrangement only allows recording the in-plane movement (radial and azimuthal direction).

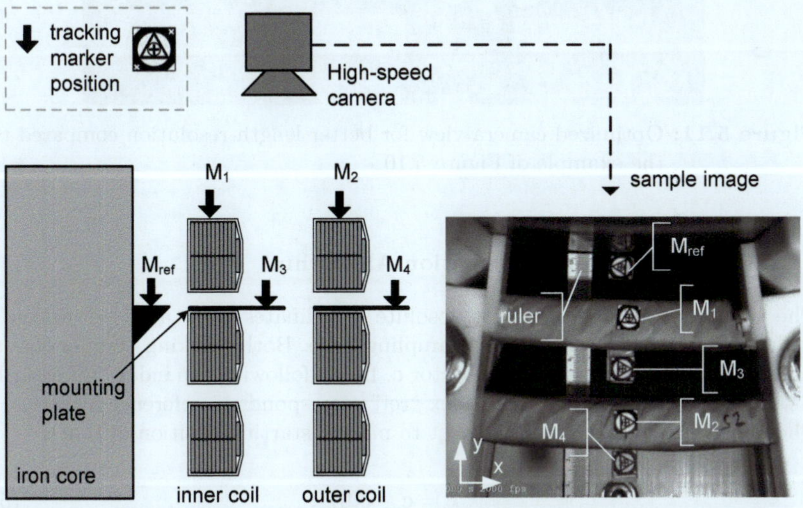

Figure 5.10: Test stand cross-sectional view (not to scale) with camera positioning, tracking markers $M_1 \dots M_4$ and reference marker M_{ref}

5.2.2.2 Marker Mounting and Image Calibration

The tracking markers are applied in the form of adhesive labels either directly on the paper insulation of the windings, or on the mounting plates. The latter are attached to the winding insulation by strong, double-sided tape.
The camera orientation has to ensure that the objective is perpendicular to the plane in which the markers are located, to avoid perspective distortion. A security distance of at least 40 cm between coil and camera prevents EMC interference coupling due to the high magnetic fields. An optical zoom objective guarantees that all markers are resolved with as many pixels as possible, and stay visible in the picture. Figure 5.11 shows a resolution-optimized camera setup. Furthermore,

a cropping of unnecessary picture parts allows increasing the camera frame rate considerably above the nominal rate.
Each re-positioning of the camera demands a calibration. For unidirectional vibrations in the radial direction, a calibration based on a linear scale close to the markers is sufficient. However, if the view frame is increased, a more elaborate 2D grid calibration is necessary. For more details about this method refer to [94].

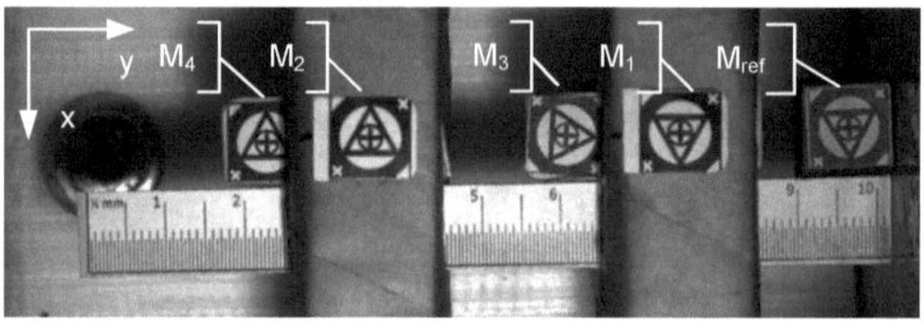

Figure 5.11: Optimized camera view for better length-resolution compared to the example of Figure 5.10

5.2.2.3 Displacement Calculation Algorithm

The tracking software estimates absolute coordinates in x- and y-direction for each marker, with respect to the sampling time. Both tracking components are summarized in a coordinate data vector **c**. In the following the index "w" describes a winding tracking marker, the index "ref" corresponds to reference marker.
The relative coordinates with respect to marker starting position $\mathbf{c}(0)$ are:

$$\tilde{\mathbf{c}} = \mathbf{c} - \mathbf{c}(0) \tag{5.5}$$

The reference marker compensates the measurements on a winding for residual test stand or camera movements. The compensated coordinate vector is given by the difference between winding and reference marker coordinates:

$$\tilde{\mathbf{c}}_{\mathrm{w,c}} = \tilde{\mathbf{c}}_{\mathrm{w}} - \tilde{\mathbf{c}}_{\mathrm{ref}} \tag{5.6}$$

If the y-axis of camera view (compare coordinate system in Figure 5.11) is perfectly aligned into the radial direction with respect to the cylindrical coils, the y-component of $\tilde{\mathbf{c}}_{\mathrm{w,c}}$ equals the radial vibration amplitude.
If the camera is misaligned, an error angle α_{e} between marker movement direction and camera y-axis arises, as visualized in Figure 5.12 a). Due to small vibration components in the angular direction, the marker trajectory becomes an ellipse, as shown in Figure 5.12 b). The angle α_{e} is given by the major axis of the best-fit ellipse. The radial movement v_{w} of a winding marker results from its y-component and α_{e} to:

$$v_w = \tilde{y}_{w,c} / \cos(\alpha_e) \tag{5.7}$$

Figure 5.13 presents a sample measurement at the middle winding of the inner coil. The red curves correspond to the measurement marker on the winding, the black curves to the reference marker, and the blue curves to the compensated data. The x-direction reveals a significant low-frequent residual movement for both winding and reference marker. This drift does not appear in the compensated signal. The final radial deformation shown in part c) reveals a pure elastic vibration, as the radial movement signal ends close to zero.

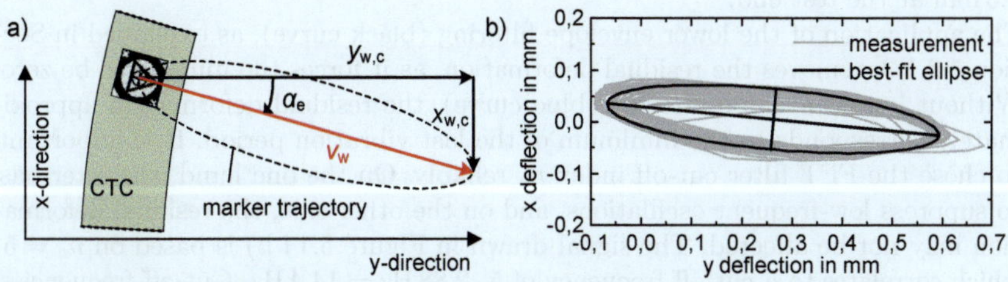

Figure 5.12: a) Schematic of misaligned camera y-axis by error angle α_e
b) Marker trajectory plot with best-fit ellipse

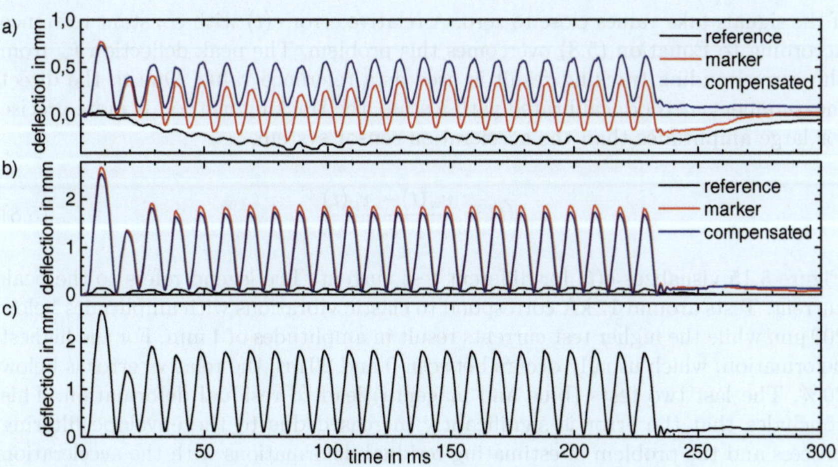

Figure 5.13: Reference and measurement marker tracking data:
a) x-component, b) y-component, c) radial component

5.2.3 Comparison of Both Systems

Tests with acceleration sensor and tracking marker applied in parallel to a test winding, facilitate a comparison of both measurement systems. The signal processing for the acceleration sensor system is based on Figure 5.8, method c). The measurement results shown in Figure 5.14 a) reveal a satisfying coincidence between the systems. Both phase and frequency of the oscillating signals coincidence. Only slight deviations in the amplitude are present.
However, Figure 5.14 b) shows a significant disadvantage of the acceleration sensors and the underlying signal processing method. The measurements correspond to a peak test current of 28.5 kA, which lead to a residual deformation of around 0.5 mm at the test end.
The application of the lower envelope filtering (black curve), as explained in Section 5.2.1.3, removes the residual deformation, as it forces the minima to be zero. Without lower envelope filtering (blue curve), the residual deformation approximately corresponds to the minimum of the last vibration period. It is important to chose the FFT filter cut-off index n_c reliably. On the one hand, the filter has to suppress low-frequent oscillations, and on the other side, the residual deformation may not be affected. The signal drawn in Figure 5.14 b) is based on $n_c = 5$, which correlates to a cut-off frequency of $5 \cdot 2.88\,\text{Hz} = 14.4\,\text{Hz}$. Cut-off frequencies between $10 \ldots 20\,\text{Hz}$ have been pointed out to be suitable.

The deviation between camera tracking result $v_w(t)$ and acceleration sensor measurement $v_g(t)$ is difficult to visualize: The relative error will increase considerably, if the signals take values close to zero. A relative error $e(t)$ with constant reference according to Equation (5.8) overcomes this problem. The peak deflection $\hat{v}_w$ from the corresponding tracking result is used as a reference value. Due to the direct measurement method, calibrated with a scale, the tracking method is more precise for large amplitudes than the acceleration sensor system.

$$e(t) = \frac{v_w(t) - v_g(t)}{\hat{v}_w} \tag{5.8}$$

Figure 5.15 visualizes $e(t)$ for different test current. The legend refers to the peak current. Tests around 12 kA correspond to elastic vibrations with amplitudes below 200 µm, while the higher test currents result in amplitudes of 1 mm. For the highest deformation, which usually occurs between 0 and 20 ms, the relative error is below 20 %. The last two tests (blue and magenta) lead to residual deformation. This concludes that the error is significantly increased due to the envelope filtering process and the problem of estimating residual deformations with the acceleration sensor system.

To conclude, each measurement system has its advantages: Acceleration sensors are cheap compared to the high-speed camera. They are easy to mount, and the data processing methods developed in this section provide a direct calculation of

displacement signals. They are most suitable for measurement of low-amplitude elastic vibrations, without residual deformation.
The camera system is more elaborate to install and calibrate. In return, it can track any number of measurement points which are easier to apply, compared to the acceleration sensors. Motion tracking analysis from the output data is more time consuming, and needs more intervention from the user. The main advantage is the possibility to measure residual deformation, which is crucial for investigations presented in this work. Therefore, the camera system is the preferred measurement method for the experiments presented in the following sections. Additional acceleration sensor measurements have been presented in [86].

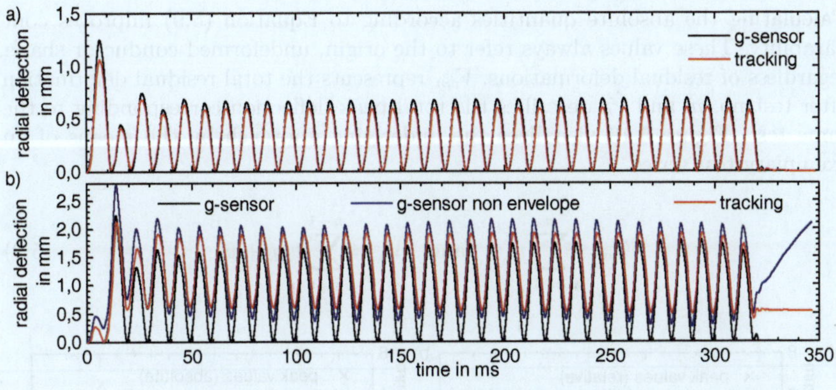

Figure 5.14: Comparison between acceleration sensor and tracking system:
a) Elastic vibration with a peak current of 20.4 kA
b) Plastic deformation corresponding to a peak current of 28.5 kA

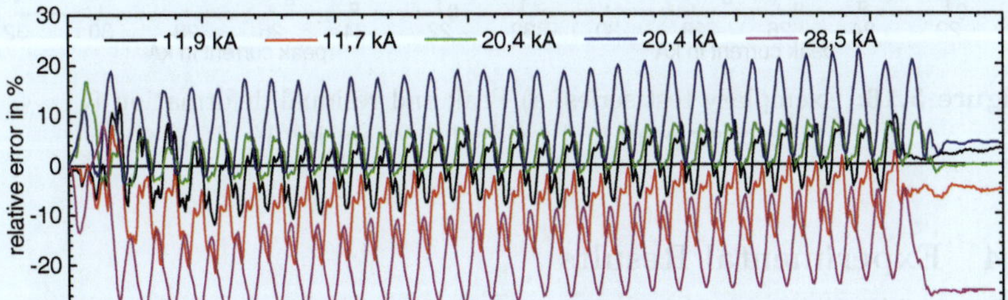

Figure 5.15: Relative error $e(t)$ between tracking and acceleration sensor measurement with envelope filtered acceleration sensor results

5.3 Measurement Data Evaluation

The most interesting points from the time-dependent deformation signal are the peak deformation $\hat{v}_n$, which is correlated to the maximum peak current, and the residual deformation $v_{\mathrm{r},n}$ at the end of a measurement with the index n. Usually between 5 ... 10 tests with rising test current are necessary, until the tested inner winding sector buckles. The labels "w" and "g", referring to the marker tracking or acceleration sensor system, are omitted from here on to improve clarity.
As long as the residual deformation is close to zero, $\hat{v}_n$ refers to the undeformed, original shape of the conductor sector under test. However, if a residual deformation occurs in test no. n, the measurement results from the subsequent test no. $n+1$ refer to the already deformed conductor.
Calculating the absolute quantities according to Equation (5.9) improves comparability. These values always refer to the origin, undeformed conductor shape, regardless of residual deformations. $V_{\mathrm{r},n}$ represents the total residual deformation after test no. n, and $\hat{V}_n$ describes the total peak deflection corresponding to the same test. Figure 5.16 visualizes the underlying approach by the means of an exemplary test series.

$$V_{\mathrm{r},n} = \sum_{k=1}^{n} v_{\mathrm{r},k} \quad \text{and} \quad \hat{V}_n = \sum_{k=1}^{n-1} v_{\mathrm{r},k} + \hat{v}_n \tag{5.9}$$

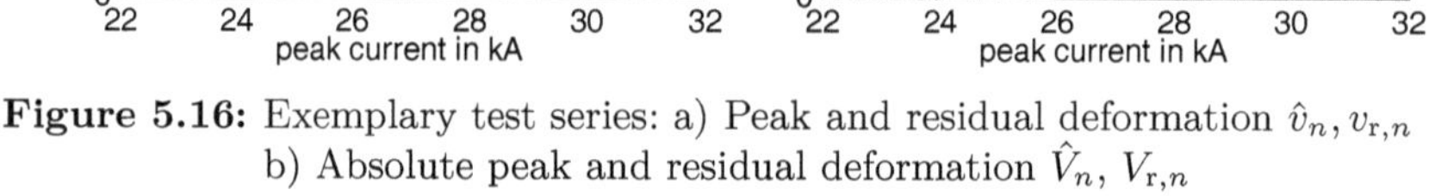

Figure 5.16: Exemplary test series: a) Peak and residual deformation $\hat{v}_n, v_{\mathrm{r},n}$
b) Absolute peak and residual deformation $\hat{V}_n$, $V_{\mathrm{r},n}$

5.4 Experimental Results

5.4.1 Testing Method Validation

The following investigation validates the testing approach based on a stepwise increase in test current, until the initiation of buckling: Four test series on independent coil sectors start at a low test current, which only causes elastic winding vibrations. The fifth test on another undeformed sector begins with a much higher

current which will initiate significant residual deformation. If the absolute deformations from the first four tests coincidence with the results from the last test, the sequential testing method is applicable.
Figure 5.17 visualizes the corresponding results for a test winding of type CTC-B-21. The measurement results from the fifth, high current test perfectly lie on the average line, which deduced from the preceding tests starting with elastic vibrations. This indicates that the proposed sequential testing method is independent of the starting current. The start with a low current enables a more precise prediction of the critical current which initiates buckling. For the example of Figure 5.17 this point is between 24 . . . 26 kA as the average residual deflection begins to take values $\geq 0.5\,\mathrm{mm}$. Due to the scattering of measurement results, the measurement of several independent winding sectors is necessary to gain precise results.

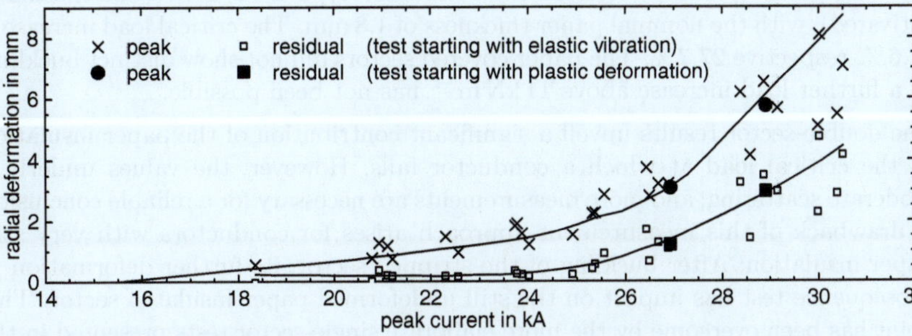

Figure 5.17: Comparison of test series starting at low currents to a test initiated at a higher current (filled markers)

5.4.2 Influence of Paper Insulation

The experiments presented in this section evaluate the impact of the paper insulation on the buckling strength of CTCs. It has not been possible to get or to produce aged conductors with a sufficient length for the test stand. To overcome this, different insulation thicknesses have been considered. A blank conductor represents the worst-case of a heavily aged insulation. The preliminary work on this topic is documented in [94].

5.4.2.1 Double-Sector Tests

This testing method regards two adjacent winding sectors in parallel. One sector remains in its original state, while the paper insulation is unwrapped from the inner winding of the other sector, as shown in Figure 5.18 a): The stripped sector represents a winding with heavily aged paper insulation, which practically is not present anymore. The high-speed camera system tracks both sectors in parallel.

Due to the wide view frame, a 2D calibration reduces optical distortion. Figure 5.18 b) shows the underlying ruler.
Figure 5.19 provides the test results for two different CTC types in the form of absolute peak and residual deformation at the middle turn of the inner coil. The data is plotted with respect to the electromagnetic load, which acts on this turn. Section 5.5.1 explains the calculation of mechanical load from test current. The limit of 0.5 mm for the residual deformation defines the critical load. Vertical, dotted lines mark the resulting values in Figure 5.19.

For the conductor type CTC-B-21 three tests have been conducted: Part a) and b) of Figure 5.19 correspond to a paper thickness of 0.9 mm. The increase in critical load is 24.6 %, respective 19.6 %. Part c) corresponds to $p = 1.8$ mm resulting in a load increase of 28.8 %.
The last parts d) and e) refer to the conductor type CTC-D-37 (not thermally activated), with the nominal paper thickness of 1.8 mm. The critical load increase is 31.6 %, respective 27.7 %. The paper covered sectors did not show distinct buckling, as a further load increase above 11 kN m^{-1} has not been possible.

The double-sector results unveil a significant contribution of the paper insulation to the critical load at which a conductor fails. However, the values underly a moderate scattering, and more measurements are necessary for a reliable conclusion. A drawback of this measurement approach arises for conductors with very stiff paper insulation: After buckling of the stripped sector, its further deformation in subsequence test has impact on the still undeformed paper insulated sector. This point has been overcome by the more elaborate single-sector tests presented in the following.

5.4.2.2 Single-Sector Tests

The measurement of only one sector per test series is more efficient concerning image calibration and reproducibility of results. Furthermore, the impact of the weak stripped sector on the stronger paper covered sector, arising in the double-sector approach, does not appear. A drawback is the increasing test duration due to two independent test series.
Several tests have been conducted with CTC-B-21 and CTC-E-19 with varying insulating thickness at the inner coil of the test stand. Due to its high stiffness, CTC-D-37 has not been considered. Figure 5.20 visualizes some exemplary camera recordings from a single-sector test series. The upper row shows the sector before the test and the bottom row at the test end. The first two tests a) and b) mainly are elastic, as no residual deformation occurs. Part c) ends with a significant residual deformation.

The experimental results conducted with CTC-B-21 consider four cases: $p = 0$ mm; 0.45 mm; 0.9 mm and 1.8 mm. Figure 5.21 a) visualizes the residual deformation results. Two clusters are visible: One for $p = 0$ mm and 0.45 mm, the second cluster exists for a paper covering of 0.9 mm and 1.8 mm. It is barely possible to recognize

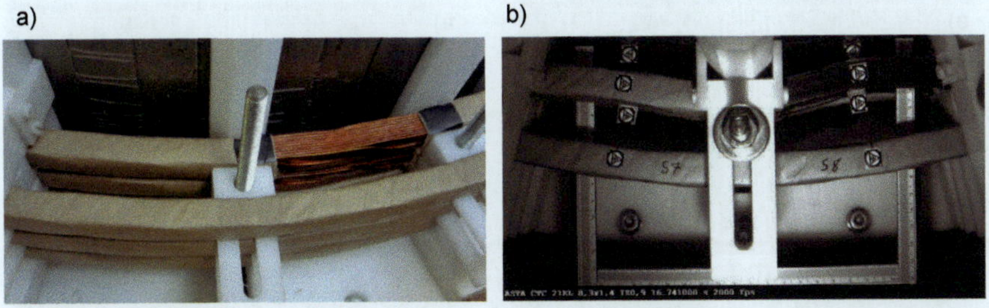

Figure 5.18: a) CTC-B-21 winding arrangement with 2 paperless upper turns at the right sector
b) Camera recording at a peak load of $16\,\mathrm{kN\,m^{-1}}$ during a test

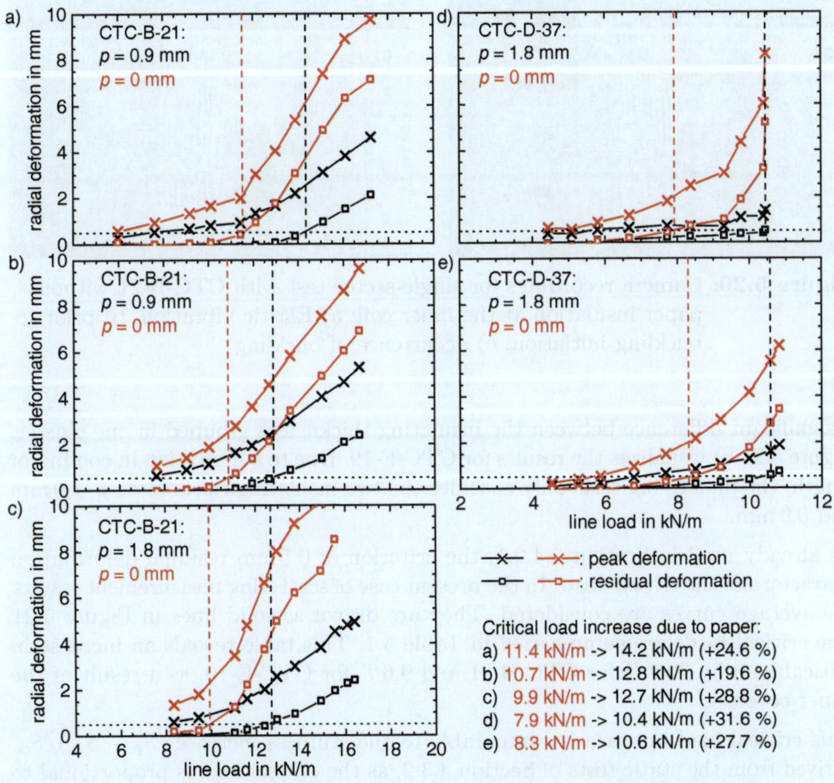

Figure 5.19: Maximum deformation resulting from dynamic double-sector tests: With paper insulation (black data) and stripped (red data)

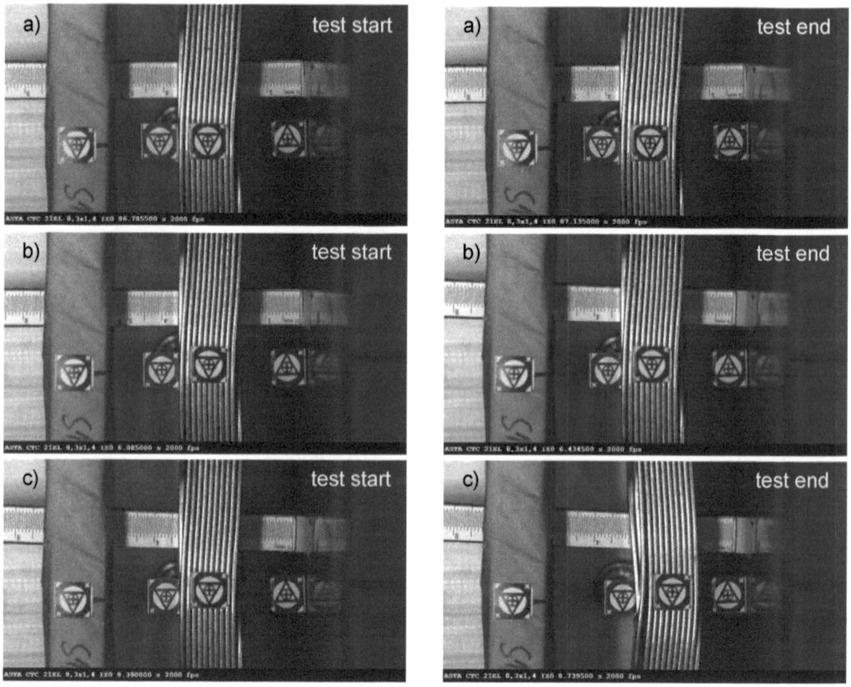

Figure 5.20: Camera recordings for single-sector test with CTC-B-21 without paper insulation at the inner coil: a) Elastic vibration, b) prior to buckling initiation, c) occurrence of buckling

a significant difference between the insulating thicknesses grouped in one cluster. Figure 5.21 b) visualizes the results for CTC-E-19. Due to a limitation in conductor length, the underlying tests only consider the two insulating thicknesses $p = 0\,\mathrm{mm}$ and 0.9 mm.

As already used in Section 5.4.2.1, the criterion of 0.5 mm residual deformation characterizes the critical load. In the present case of scattering measurement results, the average curves are considered. They are drawn as solid lines in Figure 5.21. The critical loads are summarized in Table 5.1. This table reveals an increase in critical load by 30.7 % for CTC-B-21 and 9.6 % for CTC-E-19, as a result of the paper covering.

This critical load increase is comparable to the stiffness increase $(S_{\mathrm{p}} - S_{\mathrm{co}})/S_{\mathrm{co}}$ derived from the static tests of Section 4.3.2, as the critical load is proportional to the stiffness $S = E_0 I$ for elastic material properties, as given in Equation (3.19). According to Figure 4.9 a) the stiffness increase for CTC-B-21 is 31.9 % as result of a 1.8 mm paper covering. This is quite close to the critical load increase of 30.7 %

gathered from the static tests. In the case of CTC-E-19, the static bending tests lead to a stiffness increase by 12.8 % for a paper covering of 0.8 mm. This value is once again close to the critical load increase of 9.6 %.

To conclude, the stiffness increase measured in static bending tests approximately corresponds to the increase in critical load deduced from dynamic short-circuit tests.

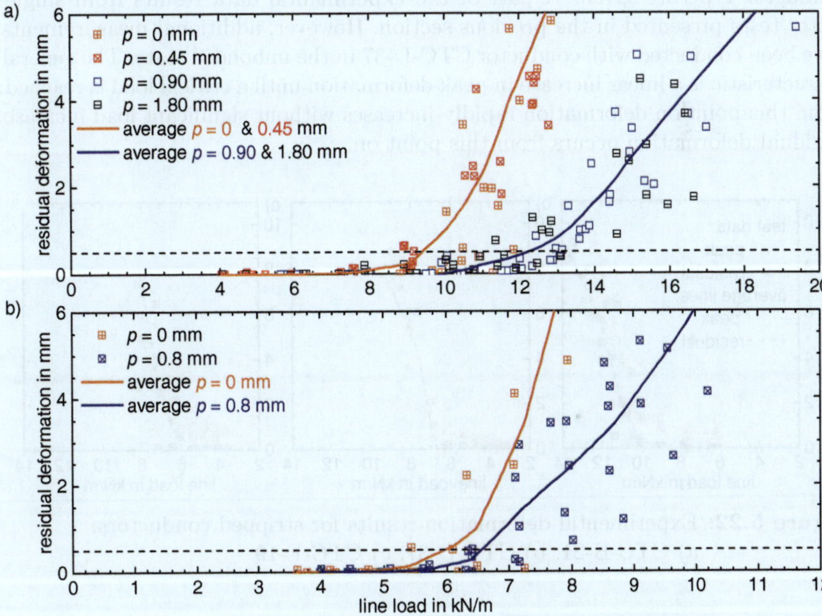

Figure 5.21: Residual deformation test results with average lines: a) CTC-B-21, b) CTC-E-19

Table 5.1: Results deduced from single sector tests

	p in mm	q_{cr} in kN m^{-1}	q_{cr} increase in %
CTC-B-21	0 … 0.45	9.39	0
	0.9 … 1.8	12.28	30.7
CTC-E-19	0	5.95	0
	0.8	6.52	9.6

5.4.3 Tests on Stripped CTCs

The experimental results on stripped CTCs serve to check their accordance with the results gathered from FEA studies or analytic calculations. The fundamental approach is to compare the peak displacement-load characteristics.
Figure 5.22 gives the results as total peak and residual deformation referring to the peak line load. Furthermore, the corresponding average curves for each conductor type are given. A part of the experimental data results from single-sector tests presented in the previous section. However, additional measurements have been conducted with conductor CTC-D-37 in the unbonded state. The general characteristic is a linear increase in peak deformation until a critical load is reached: From this point on deformation rapidly increases without significant load increase. Residual deformation occurs from this point on.

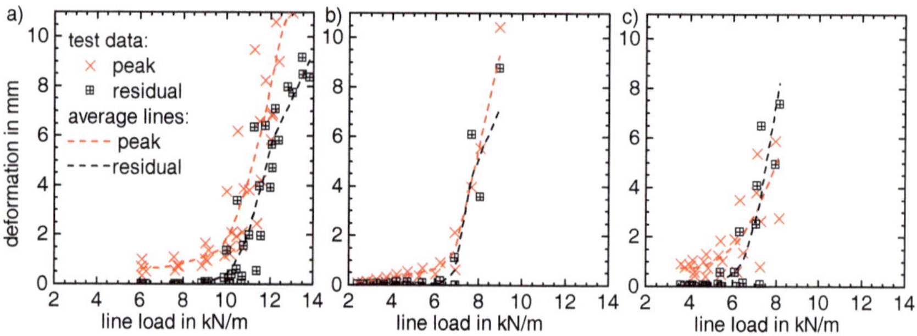

Figure 5.22: Experimental deformation results for stripped conductors: a) CTC-B-21, b) CTC-D-37, c) CTC-E-19

5.5 FEA Model

The simulation model is based on a coupled FEA study. A 2D magnetostatic analysis calculates the magnetic force distribution on the windings from the peak test current. An unidirectional coupling link maps this force density to a 3D static structural FEA, which itself calculates the deformation behavior of the windings. Preliminary investigations to the simulation model development are given with the student thesis [97]. The approaches and experimental results presented in this section have been published in a shortened summary publication [81].
Although short-circuit currents and forces are of dynamic nature, a static approach has been turned out to be sufficient for the given test stand. According to [36, p. 247], the contribution of dynamic forces is negligible for radial buckling phenomena. In conclusion, the present work neglects, like many other others, the contribution of conductor inertia with respect to the low excitation frequency of 100 Hz [6, 7].

On the electrical side, the transposed conductors and the sheeted iron core significantly reduce transient effects like eddy currents or the proximity effect. In the same way, the skin effect has no relevance. The skin depth for copper at a frequency of 50 Hz and a temperature of 60 °C lies at 9.2 mm [17, p. 461]. The radial thickness of CTC strands usually is much smaller than this value. In conclusion, the current distribution inside each strand is homogeneous and the magnetostatic approach is applicable.

5.5.1 2D Magnetostatic Analysis

A rotation symmetric 2D analysis is the simplest way to calculate the stray magnetic field between the coils wound on the test stand. Only the lead exits, near to the clamping supports, introduce some distortion. However, winding sectors near to this area are not used for testing. Figure 5.23 a) shows the interesting part of the 2D model, for a typical winding arrangement from type CTC-B-21. The outer borders, which limit the model by a balloon boundary, are not shown.
All insulating materials with a relative permeability of 1 (spacer, insulating paper) are neglected within the magnetostatic field calculation. The nonlinear $B(H)$-curve shown in Figure 5.25 a) defines the iron core. The conductor cross-section is homogenized. The peak test current $\hat{I}_{\mathrm{max}}$ is assigned to the inner and outer coil by the means of a current density $J_\varphi = \hat{I}_{\mathrm{max}}/A$ in circumferential direction. The sign is opposite between both windings, to respect their opposed series connection.
Figure 5.23 b) visualizes the force density vector field $\mathbf{f}$ inside the conductors according to Equation (2.51). The inner coil is pushed against the core, while the outer windings experience a force in positive radial direction. Furthermore, axial forces press both top and bottom windings towards the middle turns. The axial force on the inner turns are close to zero. For the present test stand the circumferential component f_φ is assumed to be zero.
For the investigation of buckling phenomena, the radial components are of main interest. The double integral according to Equation (5.10) defines the radial line load $q_{\mathrm{r},j}$, acting on winding no. j. The integrand $\mathrm{d}A_j$ defines the area, in which winding no j lies. This distinction is necessary, as the radial force density field f_r is not homogeneous.

$$q_{\mathrm{r},j} = \iint_{A_j} f_\mathrm{r}\,\mathrm{d}A_j = \iint_{A_j} J_\varphi B_z\,\mathrm{d}A_j \tag{5.10}$$

The load $q_{\mathrm{r},j}$ depends on the squared current, as both J_φ and B_z are linearly dependent from the current $\hat{I}_{\mathrm{max}}$ (Equation (2.12)). Therefore, the FEA results allow calculating a time independent load factor $k_{\mathrm{r},j}$ for each winding, which relates squared current and radial line load.

$$k_{\mathrm{r},j} = q_{\mathrm{r},j}/\hat{I}^2_{\mathrm{max}} \tag{5.11}$$

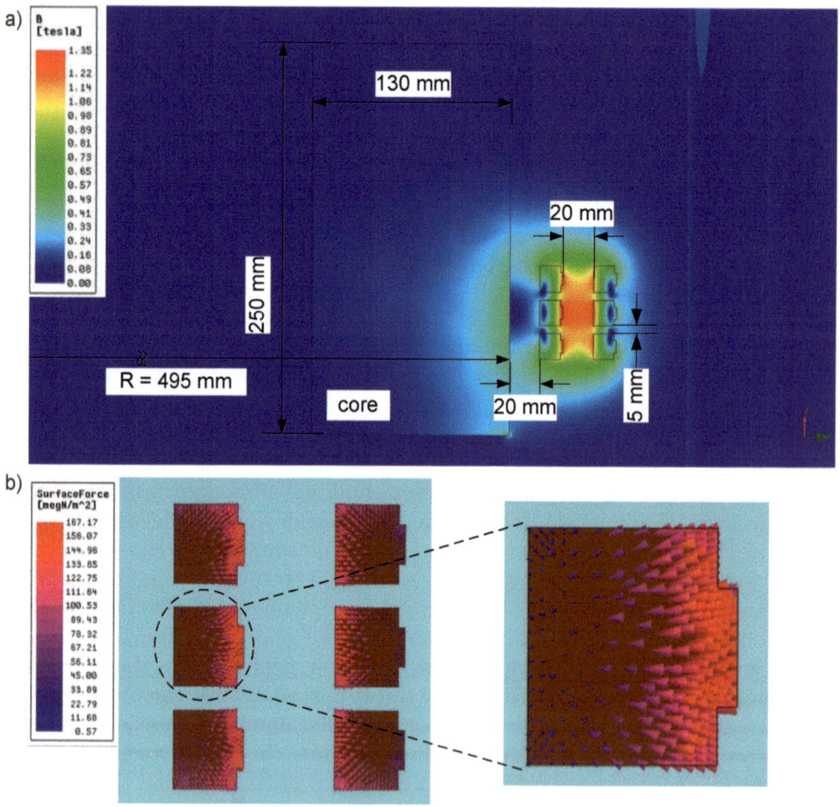

Figure 5.23: a) Cut out view of 2D test stand model for CTC-B-21 with magnetic flux in Tesla (28.5 kA peak)
b) Force density vector plot in MN/m^2 (28.5 kA peak)

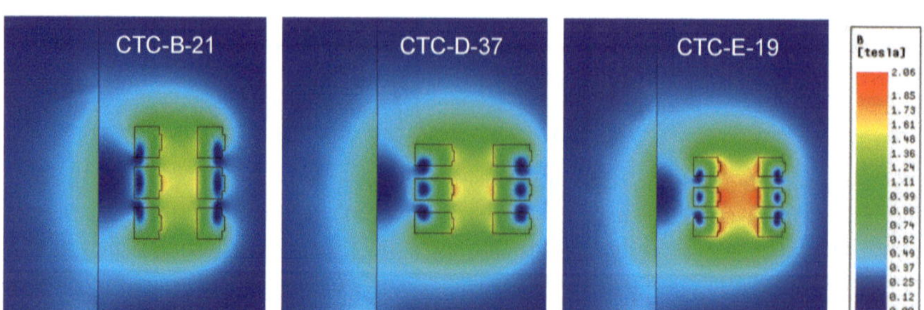

Figure 5.24: Magnetic flux density in Tesla (35 kA peak, cut out views)

With this factor the time dependent radial load $q_{\mathrm{r},j}(t)$ on conductor no. j results from the test current $i_{\mathrm{k}}(t)$ according to

$$q_{\mathrm{r},j}(t) = i_{\mathrm{k}}^2(t) \cdot k_{\mathrm{r},j}. \tag{5.12}$$

This assumption of a constant load factor is valid, as long as the core magnetization stays within the linear range of the hysteresis curve. Otherwise, the stray field outside the core does not depend linearly on the current anymore. The present test stand guarantees linearity for test currents up to 35 kA, based on the typical winding arrangements used in this work. The core saturation results presented in Figure 5.24 prove that the maximum flux inside the core of approximately 1 T stays within the linear range for a test current of 35 kA.
The three CTC types tested in this work lead to load factors according to Table 5.2. $k_{\mathrm{z},j}$ refers to the axial load factors. Index $j = 1$ corresponds to the upper turn, $j = 2$ represents the middle turn and $j = 3$ appoints to the bottom turn. The factors are not perfectly symmetric, as the winding is not axially aligned in the core height middle section. However, it is given approximately that $k_{\mathrm{r},1} \approx k_{\mathrm{r},3}$ and $|k_{\mathrm{z},1}| \approx |k_{\mathrm{z},3}|$. The middle turn experiences the largest force in the radial direction, and almost no force in the axial direction. Comparing the radial and axial factors concludes that the axial force contributions are also considerable. But axial bending is not an issue, as for CTCs $h_{\mathrm{s}} \gg t_{\mathrm{s}}$. All load factors depend on spacer dimensions, conductor cross-section and insulating thickness, and remain constant, as long as no displacements occur.

Table 5.2: Load factors for the test stand inner coil

unit: N/m/kA^2	$k_{\mathrm{r},1}$	$k_{\mathrm{r},2}$	$k_{\mathrm{r},3}$	$k_{\mathrm{z},1}$	$k_{\mathrm{z},2}$	$k_{\mathrm{z},3}$
CTC-B-21	13.45	15.67	13.33	-9.11	0.29	9.71
CTC-D-37	13.61	15.10	13.50	-11.56	0.35	12.27
CTC-E-19	16.11	18.03	16.04	-13.74	0.26	14.28

With regard on radial buckling and due to the unidirectional coupling, it is necessary to check the validity range for the constant load factors. A change in coil structure influences the magnetic field distribution and the magnetic forces on the conductors. The assumption of a rotation symmetric arrangement becomes invalid, as the conductors' radial position varies between sector middle (highest deformation) and axial stick position (zero deformation). It follows that precise post-buckling results are only possible with a fully coupled 3D analysis.
However, the main focus of this work lies on the buckling initiation. Minor deviations in the post-buckling behavior are acceptable. Nevertheless, the dependency between load factor and inner coil displacement has been analyzed by the means of

a 2D rotation symmetric FEA, to point out the impact of radial coil displacements. The following two cases have been considered:

1. All inner turns deform uniformly towards the coil middle until a maximum deflection of 5 mm is reached. This case corresponds to elastic vibration, where the deformation of all turns approximately is equal as $k_{r,1} \approx k_{r,2} \approx k_{r,3}$.
2. The middle turn's radial deformation is twice as high than the deformation at the uppermost and lowermost turns. If, for example, the middle turn is deformed by 5 mm, the deformation at the remaining inner turns is 2.5 mm. This case of misaligned deflection corresponds to buckling phenomena which always start at the middle turn, and may result in large radial misalignments at the inner coil.

Figure 5.26 visualizes an exemplary simulation result for CTC-D-37 at a test current of 35 kA peak. Part a) shows the magnetic flux distribution without radial

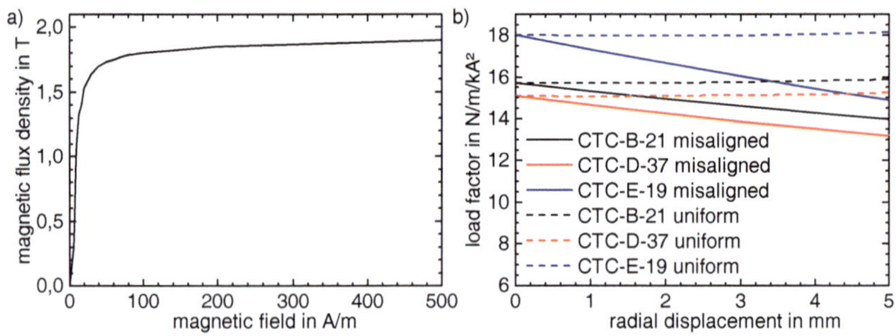

Figure 5.25: a) Iron core nonlinear magnetization curve
b) Displacement-dependent radial load factor variation for inner, middle turn

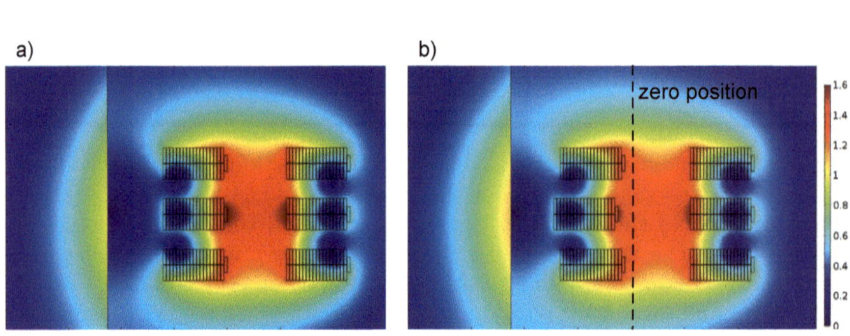

Figure 5.26: Magnetic flux density in Tesla (35 kA peak, cut out views)
a) Zero radial deformation, b) 5 mm radial deformation at middle, inner turn, 2.5 mm at remaining inner turns

displacement, whereas in part b) the radial position of the inner coil is changed due to a misaligned deformation: The middle turn is displaced by 5 mm, the remaining turns by 2.5 mm.
Figure 5.25 b) plots the radial load factors at the middle inner turn in dependency of the radial coil deformation. The factors almost remain constant for all three CTC types, if the whole inner coil deforms uniformly (dashed curves). In the case of misaligned deformation (solid curves), they decrease with increasing deflection. To conclude, for elastic vibration the unidirectional coupled FEA and constant load factors are applicable. In the case of distinct displacements with radially misaligned conductors, a bidirectional fully coupled analysis is necessary to receive more precise results.

5.5.2 3D Static Structural Analysis

Due to the numerous parts, a complete 3D simulation for the whole test stand is not feasible. Especially the wide range of contact areas leads to an enormous increase in simulation time. In conclusion, the 3D structural analysis complexity has been reduced as shown in Figure 5.27, based on the following points:

- As the middle turn of the inner coil experiences the highest radial force (Table 5.2), only this turn is considered.
- The simplified CTC model according to Section 3.2.1 is used.
- Only a cut out from the whole test stand with the sector under test in its middle (Figure 5.27 a)) is considered. The circumference of 90° facilitates strain and slippage of the winding near to the tested unsupported sector in the circumferential direction (Figure 5.27 b)). The student thesis [97] investigated the need of this angular extent. A further reduction in calculation time is possible by the use of a symmetry plane parallel to both radial and axial axis through the middle of the sector under test (Figure 5.27 c)).
- Fixed constraints at the inner surfaces of the axial sticks which are directed to the core allow neglecting the core. This assumption is valid, as the iron core has a very high stiffness. The axial sticks have reduced height, without influencing simulation results [97].
- There are no radial spacers. Stress induced due to axial forces transferred by these spacers, does not impact the radial deformation behavior [97].

The following constraints restrict the model components' degrees of freedom, as given in the real test stand:

- The inner surfaces of the axial sticks directed to the core are rigidly fixed.
- A nodal constraint prevents any axial displacement of the middle turn: The mesh nodes of the turn, which would be in contact with the excluded radial spacers, are not allowed to move in the axial direction.
- Frictionless support boundary conditions constraint the conductor ending in the circumferential direction.
- Symmetry boundary conditions apply to the winding endings.

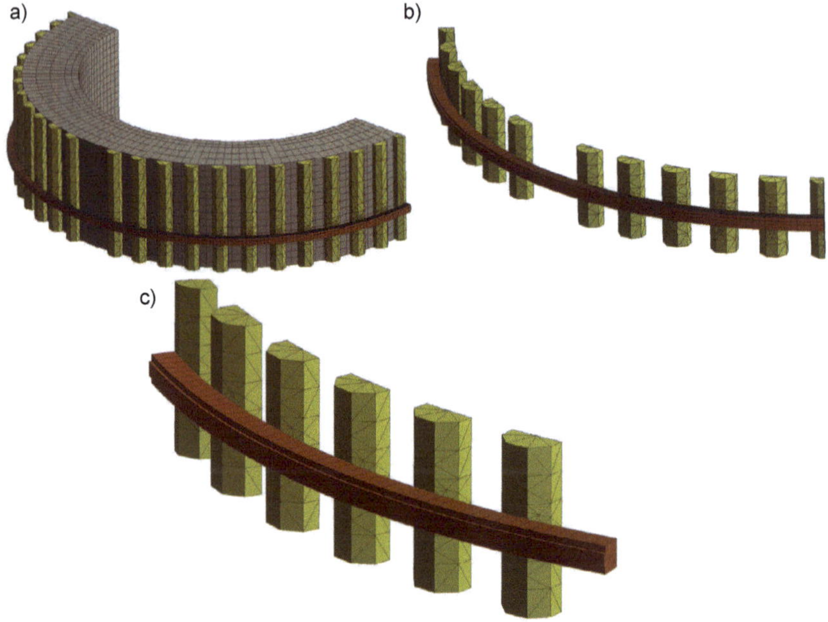

Figure 5.27: a) 180° model with core, b) 90° simplified model
c) 90° symmetry model

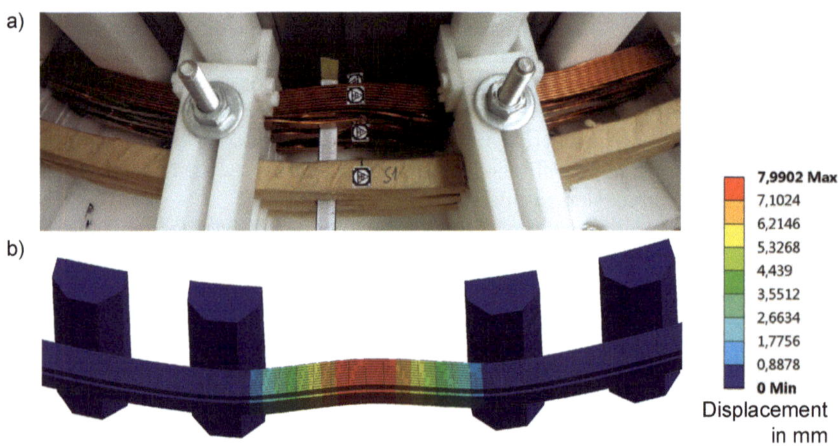

Figure 5.28: a) Deformation resulting from experiments with CTC-B-21 and b) from FEA model

Concerning mechanical property of materials, the FEA study assumes plastic spacers (POM) with linear elastic properties, based on a Young's Modulus of 2.7 GPa and a Poisson's ratio of 0.35. The copper strands have elastoplastic properties, based on a measured stress-strain characteristic (Table 4.1). The copper Poisson's ratio of 0.33 is taken from literature.
Frictional contacts define the contact areas between the strands, as all tests are conducted in a dry condition without oil impregnation. The conductors CTC-B-21 and CTC-E-19 have wax-treated enamel coated strands. The corresponding coefficient of friction given in Table B.2, has been taken from [47]. Thermosetting CTCs with unbonded adhesive coating have significantly higher values. However, there is no literature data concerning frictional coefficients of the unbonded coating. To overcome this, several simulations with varying frictional coefficients up to 0.4 have been conducted.
The coupling link between 2D magnetostatic analysis and 3D static structural model calculates a volume correlated force density from the 2D results. To respect elastoplastic effects, the 3D analysis ramps this load linearly from zero to the maximum value, and estimates the deflection characteristic for the considered turn.

5.5.3 Simulation Results

Figure 5.28 compares the final deformation resulting from the FEA in the buckling state to the experimental results. The FEA model is mirrored at the symmetry plane, to improve the clarity of results.
The deflection-load characteristics shown in Figures 5.29 to 5.31 provide a more precise comparison for the investigated conductor types. The dashed lines correspond to the experimental results from the test stand visualized in Figure 5.22. The peak deflections are given as absolute values, which are referenced to the origin, undeformed conductor position. Therefore, a comparison to the FEA results is possible.
The trend of simulated deflection-load characteristics is equal for all conductor types: A linear rise beginning in the origin until the critical load is reached. From this point on, a rapid increase in deflection without further increase in load follows. The subsequent deflection after the fast rise is limited to a value of 6 . . . 8 mm. This limitation corresponds to twice the rise of the curved conductor sector, as explained in Appendix A.3.
In the linear elastic range, the experimental peak curves perfectly match the simulated curves. However, a noticeable difference between simulation and measurement is the slope during buckling: The FEA results show a much faster rise than the experiments. Furthermore, the post-buckling deflection is higher in the tests, while the simulation results are much more limited. The results of CTC-B-21 in Fig. Figure 5.29 clearly show this deviation. For CTC-E-19 (Figure 5.30), the test current has not been increased much more than up to the initiation of buckling.

Figure 5.31 shows the influence of large frictional coefficients k_{f} for the case of the unbonded thermosetting CTC-D-37: The main difference between the simulative results is the critical buckling load which rises with increasing k_{f}. Both the linear elastic range and the post-buckling are not significantly affected by k_{f}. The linear elastic range is equal between both simulation and experimental results. The best match between measured peak average curve and simulation data is reached for $k_{\mathrm{f}} = 0.3$.

The flattened trend of the experimental peak curves during the buckling process results due to these reasons:

- Inertial effects, which are not included in the static simulation, may slow down the transition between undeformed and buckled state of the windings.
- The sequential testing method also has a slight impact on this transition.
- The FEA assumes a perfectly curved conductor. In reality, CTC transpositions and minor inaccuracies during winding process introduce imperfections, which itself defuse buckling phenomena.

The deviation at high loads after buckling results from:

- Thermal expansion of conductors due temperature increase during the tests. An elongation facilitates a further bending of the sector under test due to circumferential slippage.
- Settling and slippage effects at sticks and clamping supports.

However, the post-buckling behavior is not in the focus for design and construction of transformer windings. The focus of this work is the calculation of the critical load which initiates buckling phenomena, due to its importance for transformer engineering.
The criteria of a residual deformation above 0.5 mm defines the experimental critical buckling load. For the FEA studies, this point is characterized by the steep increase in deformation. Table 5.3 compares the experimental values to the FEA results showing a very close coincidence. However, deviations slightly increase with a change in the criteria of the experimental, critical load.

Table 5.3: Critical loads deduced from FEA model and experiments

	FEA	experiments
CTC-B-21	$9.44\,\mathrm{kN\,m^{-1}}$	$10.12\,\mathrm{kN\,m^{-1}}$
CTC-D-37	$6.83\,\mathrm{kN\,m^{-1}}$	$7.00\,\mathrm{kN\,m^{-1}}$
CTC-E-19	$6.01\,\mathrm{kN\,m^{-1}}$	$5.93\,\mathrm{kN\,m^{-1}}$

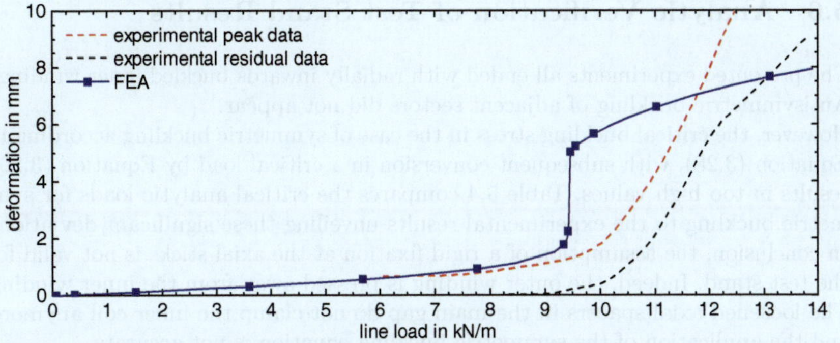

Figure 5.29: Experimental and simulation results for CTC-B-21

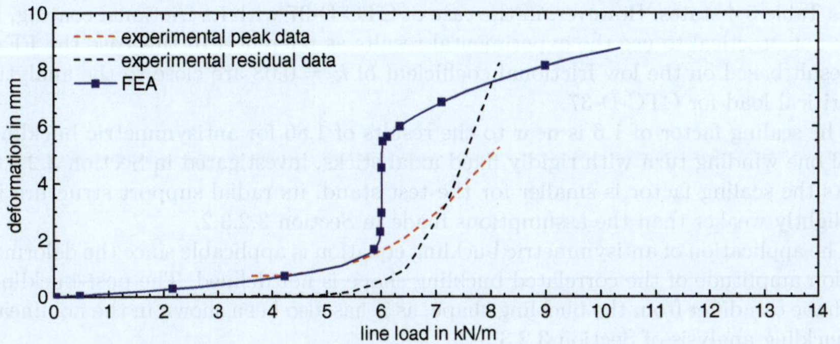

Figure 5.30: Experimental and simulation results for CTC-E-19

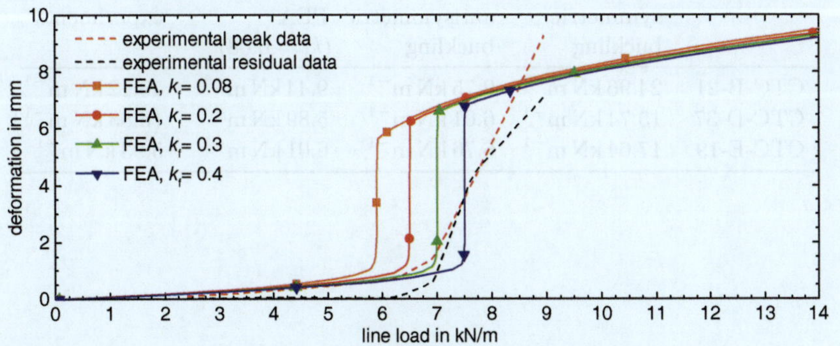

Figure 5.31: Experimental and simulation results for CTC-D-37

5.6 Analytic Verification of Test Stand Results

The presented experiments all ended with radially inwards buckled inner windings. Antisymmetric buckling of adjacent sectors did not appear.
However, the critical buckling stress in the case of symmetric buckling according to Equation (3.28), with subsequent conversion in a critical load by Equation (3.20), results in too high values. Table 5.4 compares the critical analytic loads for symmetric buckling to the experimental results unveiling these significant deviations. In conclusion, the assumption of a rigid fixation at the axial sticks is not valid for the test stand. Indeed, the outer winding is pressed away from the inner winding. The loosened axial spacers in the main gap do not clamp the inner coil anymore, and the application of the symmetric buckling equation is not accurate.
The use of the antisymmetric buckling formulation according to Equation (3.22), with scaled slenderness ratio as defined in Equation (3.59), is more promising. A scaling factor of $k_{\mathrm{sl}} = 1.6$ results in good coincidence with the experimental data, as Table 5.4 states. However, in the case of CTC-D-37 with its frictional coating, it is not practical to use the experimental results as reference. In this case the FEA result based on the low frictional coefficient of $k_{\mathrm{f}} = 0.08$ are close to the analytic critical load for CTC-D-37.
The scaling factor of 1.6 is near to the results of 1.86 for antisymmetric buckling of one winding turn with rigidly fixed axial sticks, investigated in Section 3.2.3.2. As the scaling factor is smaller for the test stand, its radial support structure is slightly weaker than the assumptions made in Section 3.2.3.2.
The application of antisymmetric buckling equation is applicable since the deformation amplitude of the correlated buckling shape is not defined. The post-buckling shape can differ from the buckling shape, as it has also been shown in the nonlinear buckling analysis of Section 3.2.3.3.

Table 5.4: Critical loads gathered from analytic consideration and FEA model

	symmetric buckling	antisymmetric buckling	FEA ($k_{\mathrm{f}} = 0.08$)	experiments
CTC-B-21	$24.96\,\mathrm{kN\,m^{-1}}$	$9.75\,\mathrm{kN\,m^{-1}}$	$9.44\,\mathrm{kN\,m^{-1}}$	$10.12\,\mathrm{kN\,m^{-1}}$
CTC-D-37	$15.74\,\mathrm{kN\,m^{-1}}$	$6.01\,\mathrm{kN\,m^{-1}}$	$5.89\,\mathrm{kN\,m^{-1}}$	$(6.83\,\mathrm{kN\,m^{-1}})$
CTC-E-19	$17.64\,\mathrm{kN\,m^{-1}}$	$6.76\,\mathrm{kN\,m^{-1}}$	$6.01\,\mathrm{kN\,m^{-1}}$	$5.93\,\mathrm{kN\,m^{-1}}$

Chapter 6

Conclusion

One focus of this work is on the mechanical characterization of Continuously Transposed Conductors (CTCs) with respect to their radial buckling strength. Experiments and Finite-Element-based studies proved, that the transposition of strands does not contribute to the conductor stiffness. Therefore, it is possible to treat a CTC as a stack of multiple strands. The critical buckling stress for a CTC without paper insulation dependents on the single strand thickness in the radial direction, as a main parameter. As long as the supporting spacers are considered as hinged supports, the analytic and simulation results are coincident, even for elastoplastic material properties. The assumption of using twice the strand thickness as equivalent conductor thickness proposed in [17, p. 275], results in too high values. However, the factor of 2 proposed in [17] approximately is applicable, if the stiffness of the radial supporting spacers is included into the analysis. Finite-Element studies based on hingeless supported windings reveal a correction factor of 1.86 for the slenderness ratio as a main parameter for the analytic equations. Since the slenderness ratio is proportional to the strand thickness in the radial direction, this factor also can be interpreted as a strand thickness scaling factor. The experimental critical loads gathered from the test stand, match to the analytic values for a correction factor of 1.6. Finally, the correction factor depends on the rigidity and dimensions of the spacers, and has to be estimated by the means of simulations.

Concerning thermosetting CTCs the adhesive coating has been characterized based on shear- and T-peel tests on bonded strands. The resulting simulation model facilitates a precise investigation of critical loads, which lead to delamination, and finally buckling of the winding. The analytical buckling equation including elastoplastic material offers a safe-side estimation for a large number of strands, since the resulting critical loads are too small. The underlying assumption is an infinite strong bonding or a homogeneous conductor cross-section with same outer dimensions as the CTC. However, it is important to ensure whether delamination is responsible for the initiation of buckling, or yielding inside the copper. Buckling due to instability is largely impossible due to the high radial CTC widths.

Another topic is the impact of the paper insulation on the overall CTC stiffness. An insulation increase of 0.9 mm leads to a significant rise in bending stiffness by up to 100 %. This stiffness increase, deduced from static bending tests, correlates with an increase in buckling strength by approximately the same percentage, as further investigations based on a dynamic short-circuit forces test stand showed. However, the stiffness increase does not linearly rise with the thickness of the overall paper covering. An insulation increase of 1 mm pointed out to be the limiting point. Furthermore, the CTC dimensions greatly influence the paper stiffness contribution. Even the approach of a homogenized Young's modulus for the paper covering resulted in CTC geometry- and insulation thickness-dependent values between 0.3 . . . 2 GPa.
Bending tests on accelerated aged CTC samples address the impact of aging. A reliable stiffness evaluation has not been possible due sticking between the strands caused by aging products. However, even a heavily aged insulation with $DP \approx 210 \ldots 280$ still contributes to the overall conductor stiffness. Compared to the factory new case, the critical deflection which caused visible and measurable paper damage, has been significantly reduced. For a more precise analysis of paper rupture, a simulation-based approach based on the aging dependent elastoplastic properties of the insulating paper is presented.

A significant part was dedicated to the experiments on the radial buckling strength of transformer windings. A test stand has been developed which allows investigating the buckling strength of test windings. The nature of electromagnetic forces is similar to the case of a real transformer. Two proposed deformation measurement systems allow estimating radial winding deformations during short-circuit current. The acceleration sensor-based approach is suitable for elastic winding vibrations. For buckling with large amplitudes and residual deformations, a high-speed camera-based system is the preferred method. The experiments unveiled that the first half-wave of the short-circuit current, which usually has the highest amplitude within ohmic-inductive systems, decides whether the winding buckles or not. If no buckling occurs, the subsequent steady-state current only causes elastic vibrations. The sequential test method based on a stepwise increased test current, slightly flattens this sharp transition between original and buckled state. However, the comparison of critical loads to the values gathered from FEA studies and analytic calculations resulted in a reliable coincidence.
For the design of short-circuit prove transformer windings it is essential to respect at least the following design parameters: The elastoplastic copper properties, the conductor cross-sectional dimensions, the radius and the number of radial supporting spacers. As the highest radial compressive forces arise on the outer turns of the low voltage winding the corresponding radius represents the worst-case. Furthermore, the slenderness ratio is lowest for the outer turns. Based on the mentioned parameters, the basic equation for radial buckling phenomena estimates a safe side

critical load with the spacers as punctual, hinged supports. If this load is higher than the electromagnetic load deduced from electromagnetic field calculations the transformer is short-circuit-proof against radial failures.

Future Work

The buckling strength of thermosetting CTCs has only been analyzed by the means of static bending tests. Dynamic short-circuit tests using the current test stand configuration have not been carried out, due to limitations in the test current amplitude. To initiate buckling, either the number of axial spacers has to be reduced, or their stiffness. The latter would correspond to a free supporting winding. Furthermore, the thermal activation of the bonding is very challenging. One possibility is to use ohmic heating, caused by a low test current through the windings. Another option is to use a large climatic chamber for heating up the whole test stand. In the case of ohmic heating, a modification of the electrical contact to the test windings is necessary, as the present clamping supports have been designed for high currents coupled with short test durations. A long term current flow will heat up the clamping supports, and electrical burning between copper and steel might occur. More work is also possible concerning the adhesive characterization. The proposed investigations have been limited to epoxy coated CTC strands. Due to their low stiffness, the copper properties contribute to the measurement results in the shear- and T-peel-tests. If it is possible to receive the coating separately, more precise tests are possible based on stiffer adherents made from steel. Some optimization work is also needed concerning the strength parameters of the bonding in the normal direction.

With regard on the stiffness contribution of the paper insulation additional CTC types should be investigated. Due to a limitation in test material the conclusions derived from this work are a first indicator, but correlations to the CTC geometry and number of layers have not been possible. Further analytic and simulation-based investigations are useful with respect on the saturation effect of the paper covering: For some conductors the stiffness decreased if the number of layers exceeded a certain limit.

Finally, it is possible to extend the present FEA studies for buckling analysis by effects like winding pretension or axial clamping forces. Especially the latter is a uncertain parameter in transformer construction, but it introduces some kind of imperfection to the windings. In contrast to this work having its focus on perfectly curved windings, it is possible to investigate the impact of imperfections on the buckling strength.

Appendix A

Analytic Derivations

A.1 Composed Cross-Sections Under Pure Bending

A beam subjected to pure bending contains a surface parallel to its upper and lower surface, where the stresses and strains are zero. With respect to the beam cross-section, this so-called neutral surface passes through the centroid of the section, as shown in Figure A.1. The resulting line is called neutral axis.

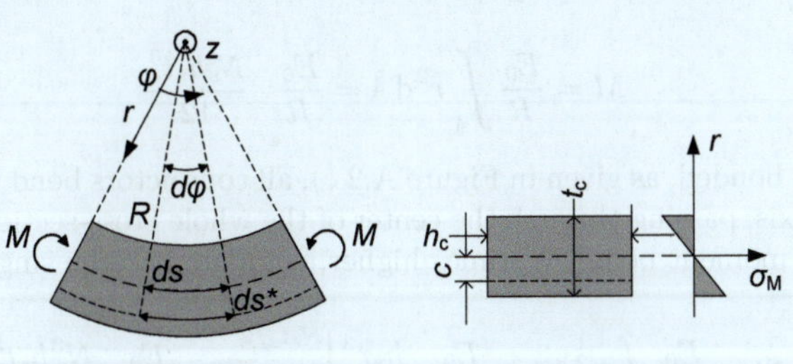

Figure A.1: Deformation of a beam in pure bending around the z-axis

Considering a beam element of $d\varphi$, the relationship between radius R (of neutral surface) and length of the neutral surface ds is

$$ds = Rd\varphi. \tag{A.1}$$

The bending strain inside a surface in the radial distance c from the neutral surface is defined by the length increase $ds^* - ds$ according to

$$\epsilon_{\mathrm{M}}(c) = \frac{ds^* - ds}{ds} = \frac{(R+c)d\varphi - Rd\varphi}{Rd\varphi} = \frac{c}{R}. \tag{A.2}$$

Using Hooke's law calculates the bending stress to

$$\sigma_{\mathrm{M}}(c) = E_0 \epsilon_{\mathrm{M}}(c) = cE_0/R. \tag{A.3}$$

Positive stress refers to tensile stress, while negative values correspond to compressive stress. The general correlation between bending stress σ_{M}, bending moment M and beam cross-section A derives from the moment balance [11, p. 229] to

$$M = \int_A r\sigma_{\mathrm{M}}(r)\mathrm{d}A, \tag{A.4}$$

with the coordinate r in the radial direction.

In the case of a rectangular shaped single conductor, as given in Figure A.2 a), Equation (A.4) results in

$$M = \frac{E_0}{R}\int_A r^2 \mathrm{d}A = \frac{E_0}{R} \cdot \left.\frac{h_{\mathrm{s}} r^3}{3}\right|_{-t_{\mathrm{s}}/2}^{t_{\mathrm{s}}/2} = \frac{E_0}{R} \cdot \frac{h_{\mathrm{s}} t_{\mathrm{s}}^3}{12} = \frac{E_0 I}{R} = \frac{S_{\mathrm{co}}}{R}, \tag{A.5}$$

by substituting $\sigma_{\mathrm{M}}(c)$ from Equation (A.3). The fraction $h_{\mathrm{s}} t_{\mathrm{s}}^3/12$ corresponds to the area moment of inertia I, with respect to the centroidal axis. The multiplication with the Young's modulus leads to the bending stiffness $E_0 I = S_{\mathrm{co}}$.

In the case of a stack of conductors, which are not bonded (Figure A.2 b)), each conductor bends around its own neutral axis. In conclusion, the bending moment of a stack of N_{s} conductors is N_{s}-times higher, compared to the single conductor:

$$M = \frac{E_0}{R}\int_A r^2 \mathrm{d}A = \frac{E_0}{R} \cdot \frac{N_{\mathrm{s}} h_{\mathrm{s}} t_{\mathrm{s}}^3}{12} \tag{A.6}$$

If the stack is bonded, as given in Figure A.2 c), all conductors bend with respect to one neutral axis, passing through the center of the whole cross-section. This results in a bending moment being N_{s}^3-times higher, compared to the single conductor:

$$M = \frac{E_0}{R}\int_A r^2 \mathrm{d}A = \frac{E_0}{R} \cdot \left.\frac{h_{\mathrm{s}} r^3}{3}\right|_{-N_{\mathrm{s}} t_{\mathrm{s}}/2}^{N_{\mathrm{s}} t_{\mathrm{s}}/2} = \frac{E_0}{R} \cdot \frac{N_{\mathrm{s}}^3 h_{\mathrm{s}} t_{\mathrm{s}}^3}{12} \tag{A.7}$$

For a bonded CTC, consisting of 2 stacks located beside each other, the axial height is $2h_{\mathrm{s}}$ and the radial thickness is $N_{\mathrm{s}} t_{\mathrm{s}}/2$. The resulting bending moment is

$$M = \frac{E_0}{R}\int_A r^2 \mathrm{d}A = \frac{E_0}{R} \cdot \left.\frac{2h_{\mathrm{s}} r^3}{3}\right|_{-N_{\mathrm{s}} t_{\mathrm{s}}/4}^{N_{\mathrm{s}} t_{\mathrm{s}}/4} = \frac{E_0}{R} \cdot \frac{N_{\mathrm{s}}^3 h_{\mathrm{s}} t_{\mathrm{s}}^3}{48}. \tag{A.8}$$

In the case of a paper insulated conductor, as shown in Figure A.2 d), both conductor and insulation bend with respect to the same centered neutral axis. Assuming a paper Young's modulus of E_{pap}, and a copper Young's modulus of E_0, leads to the bending moment according to

$$M = \frac{E_0}{R}\int_{A_{\mathrm{co}}} r^2 \mathrm{d}A_{\mathrm{co}} + \frac{E_{\mathrm{pap}}}{R}\int_{A_{\mathrm{pap}}} r^2 \mathrm{d}A_{\mathrm{pap}} \tag{A.9}$$

$$= \frac{E_0}{R}\cdot\frac{h_{\mathrm{c}}t_{\mathrm{c}}^3}{12} + \frac{E_{\mathrm{pap}}}{R}\cdot\frac{(h_{\mathrm{c}}+p)(t_{\mathrm{c}}+p)^3 - h_{\mathrm{c}}t_{\mathrm{c}}^3}{12} = \frac{1}{R}(S_{\mathrm{co}}+S_{\mathrm{pap}}). \tag{A.10}$$

This concludes that it is possible to compose the total conductor stiffness from the copper and the paper stiffness. In the case of a CTC the copper stiffness refers to $S_{\mathrm{co}} = N_{\mathrm{s}}h_{\mathrm{s}}t_{\mathrm{s}}/12$.

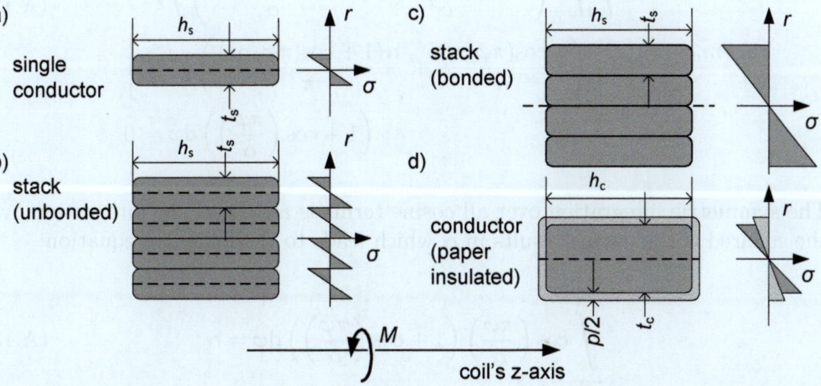

Figure A.2: Stress distribution for different conductor cross-sections under pure bending around the z-axis

A.2 Integration of Bubnov-Galerkin Integral

Substituting the differential operator Ł from Equation (3.35) into the Bubnov-Galerkin integral according to Equation (3.36) gives:

$$\begin{aligned}\int_{-\alpha}^{\alpha} \text{Ł}[u]u(t,\varphi)\mathrm{d}\varphi = \int_{-\alpha}^{\alpha} \Bigg[&\frac{\partial^6 u(t,\varphi)}{\partial\varphi^6} + 2\frac{\partial^4 u(t,\varphi)}{\partial\varphi^4} + \frac{\partial^2 u(t,\varphi)}{\partial\varphi^2} \\ &+ \frac{q_{\mathrm{r}}(t)R^3}{E_0 I}\left(\frac{\partial^4 u(t,\varphi)}{\partial\varphi^4} + \frac{\partial^2 u(t,\varphi)}{\partial\varphi^2}\right) \\ + \frac{(m_{\mathrm{c}}+k_{\mathrm{m}})R^4}{E_0 I}\frac{\partial^2}{\partial t^2}&\left(\frac{\partial^2 u(t,\varphi)}{\partial\varphi^2} - u(t,\varphi)\right)\Bigg] u(t,\varphi)\mathrm{d}\varphi \stackrel{!}{=} 0\end{aligned} \tag{A.11}$$

The substitution of $u(t,\varphi)$ according to Equation (3.37) leads to Equation (A.12). The factor $f(t)\cdot\alpha/\pi$ has been cut from the last $u(t,\varphi)$ outside Ł$[u]$, based on the assumption that it is not zero.

$$\begin{aligned}\int_{-\alpha}^{\alpha}\Bigg[\Bigg(&\frac{\pi^5\cos(\pi\varphi/\alpha)}{\alpha^5}-2\frac{\pi^3\cos(\pi\varphi/\alpha)}{\alpha^3}+\frac{\pi\cos(\pi\varphi/\alpha)}{\alpha}\\&+\frac{q_{\mathrm{r}}(t)R^3}{E_0I}\left(-\frac{\pi^3\cos(\pi\varphi/\alpha)}{\alpha^3}+\frac{\pi\cos(\pi\varphi/\alpha)}{\alpha}\right)\Bigg)f(t)\\+\frac{(m_{\mathrm{c}}+k_{\mathrm{m}})R^4}{E_0I}&\left(\frac{\pi\cos(\pi\varphi/\alpha)}{\alpha}+\frac{\alpha(1+\cos(\pi\varphi/\alpha))}{\pi}\right)f''(t)\Bigg]\\&\cdot\left(1+\cos\left(\frac{\pi\varphi}{\alpha}\right)\right)\mathrm{d}\varphi\overset{!}{=}0\end{aligned}\tag{A.12}$$

The symmetric integration over all cosine terms is zero, and the integration over the squared cosine terms results in α which leads to the following equation:

$$\int_{-\alpha}^{\alpha}\cos\left(\frac{\pi\varphi}{\alpha}\right)\left(1+\cos\left(\frac{\pi\varphi}{\alpha}\right)\right)\mathrm{d}\varphi=\alpha\tag{A.13}$$

Applying Equation (A.13) equations to Equation (A.12) solves the integral to

$$\begin{aligned}\left(\frac{\pi^5}{\alpha^4}-\frac{2\pi^3}{\alpha^2}+\pi+\frac{q_{\mathrm{r}}(t)R^3}{E_0I}\left(-\frac{\pi^3}{\alpha^2}+\pi\right)\right)f(t)\\+\frac{(m_{\mathrm{c}}+k_{\mathrm{m}})R^4}{E_0I}\left(\pi+\frac{3\alpha^2}{\pi}\right)f''(t)\overset{!}{=}0.\end{aligned}\tag{A.14}$$

The multiplication of Equation (A.14) by π/α^2 gives

$$\begin{aligned}\left(\frac{\pi^6}{\alpha^6}-\frac{2\pi^4}{\alpha^4}+\frac{\pi^2}{\alpha^2}+\frac{q_{\mathrm{r}}(t)R^3}{E_0I}\left(-\frac{\pi^4}{\alpha^4}+\frac{\pi^2}{\alpha^2}\right)\right)f(t)\\+\frac{(m_{\mathrm{c}}+k_{\mathrm{m}})R^4}{E_0I}\left(\frac{\pi^2}{\alpha^2}+3\right)f''(t)\overset{!}{=}0.\end{aligned}\tag{A.15}$$

The substitution of $\pi/\alpha=\xi$ results in Equation (3.38).

A.3 Rise and Span of a Winding Sector

In the case of a curved winding sector as visualized in Figure A.3, the rise h_r and the span L are defined according to

$$h_r = R\left(1 - \cos\left(\frac{\alpha}{2}\right)\right) \approx \frac{R\alpha^2}{8} \tag{A.16}$$

$$L = 2R\sin\left(\frac{\alpha}{2}\right). \tag{A.17}$$

The approximation in rise bases on the correlation $\cos\alpha \approx 1 - \alpha^2/2$. This approximation is valid even for relatively large angles: For $\alpha = 45° \hat{=} 0.785\,\mathrm{rad}$, $\cos(0.785) = 0.707$ and $1 - 0.785^2/2 = 0.692$.
The authors of [58] defined a dimensionless rise parameter as the ratio between rise to one-half of the thickness arch thickness. Written in the means of the given parameters leads to the dimensionless rise h_r^* according to:

$$h_r^* = \frac{h_r}{t_s/2} \approx \frac{R\alpha^2}{4t_s} = \frac{\pi^2 R}{t_s N_{as}^2} \tag{A.18}$$

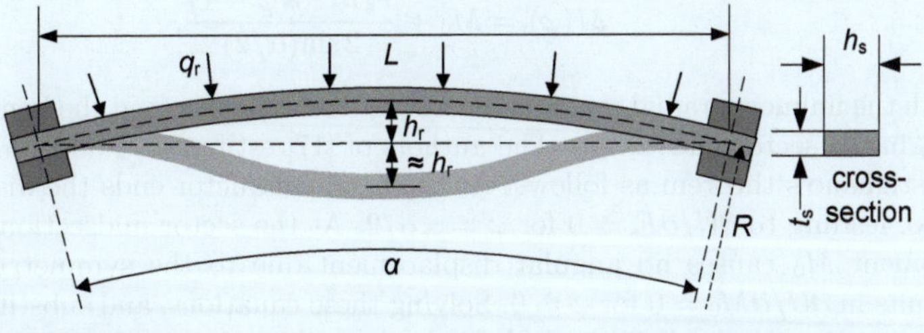

Figure A.3: Winding Sector with fixed ends under a radial line load

A.4 Winding Sector Under a Radial Compressive Load

The present approach has been derived from [17, p. 291]. This book describes the bending stress inside a curved beam under the influence of a radial compressive load. The authors transfer this problem to a winding sector supported by two spacers, and use an energy based method for the statically indeterminate structural problem. The underlying method based on Castigliano's second theorem is valid for linear elastic structures. The basic principle is, that a change in energy stored in the structure is equal to the product from one half times causing force times resulting displacement. The latter product is the fundamental definition of the energy stored in a linear elastic body.

In contrast to the approach [17, p. 291], the following considerations neglect the energy stored in the spacers. Furthermore, additional equations calculate both the conductor deflection and the strain distribution.
The considered structure is shown in Figure A.3 with conductor thickness t_s, height h_s, bending radius R and line load q_r. The sector extends from $\varphi = -\alpha/2 \ldots \alpha/2$, corresponding to the marked angle α. The spacers act as rigid fixations at $\varphi = \pm\alpha/2$.
The total strain energy U inside the conductor is

$$U = 2\int_0^{\alpha/2} \frac{F_n^2(\varphi)R}{2E_0 A}\mathrm{d}\varphi + 2\int_0^{\alpha/2} \frac{M^2(\varphi)R}{2E_0 I}\mathrm{d}\varphi, \tag{A.19}$$

with normal force $F_n(\varphi)$, Young's modulus E_0, cross-sectional surface $A = h_s t_s$, bending moment $M(\varphi)$ and area moment of inertia $I = h_s t_s^3/12$. The first integral corresponds to the energy associated with tensile or compressive forces, while the second integral describes the energy caused by the bending moment. The unknown functions $F_n(\varphi)$ and $M(\varphi)$ are given by

$$F_n(\varphi) = \frac{F_s \cos\varphi}{2\sin(\alpha/2)} - q_r R \tag{A.20}$$

$$M(\varphi) = M_0 + \frac{F_s R(\cos\varphi - 1)}{2\sin(\alpha/2)}, \tag{A.21}$$

with the unknown radial reaction force F_s at the supports, and the bending moment M_0 in the sector midsection. The authors of [17] estimate F_s and M_0 by applying Castigliano's theorem as follows: At the fixed conductor ends the displacement is zero, leading to $\partial U/\partial F_s = 0$ for $\varphi = \pm\alpha/2$. At the sector midsection the bending moment M_0 causes no angular displacement due to the symmetry. The latter results in $\partial U/\partial M_0 = 0$ for $\varphi = 0$. Solving these equations, and substituting F_s and M_0 into Equations (A.20) and (A.21) leads to the unknown functions according to

$$F_n(\varphi) = q_r R\left[\frac{\cos\varphi}{2\sin(\alpha/2)}\left[\frac{1}{g_1(\alpha) + g_2(\alpha)\cdot AR^2/I}\right] - 1\right] \tag{A.22}$$

$$M(\varphi) = \frac{q_r R^2}{2}\left[\frac{\cos\varphi}{\sin(\alpha/2)} - \frac{1}{\alpha/2}\right]\cdot\left[\frac{1}{g_1(\alpha) + g_2(\alpha)\cdot AR^2/I}\right], \tag{A.23}$$

with the two functions $g_1(\alpha)$ and $g_2(\alpha)$ given by

$$g_1(\alpha) = \frac{1}{4\sin^2(\alpha/2)}\left(\frac{\alpha}{2} + \frac{\sin\alpha}{2}\right) \tag{A.24}$$

$$g_2(\alpha) = \frac{1}{4\sin^2(\alpha/2)}\left(\frac{\alpha}{2} + \frac{\sin\alpha}{2}\right) - \frac{1}{\alpha}. \tag{A.25}$$

According to [8, p. 100] the radial deflection $v(\varphi)$ derives from the bending moment $M(\varphi)$. For $\alpha \ll 1$, so that $\arctan\alpha \approx \alpha$, the curvature $-1/R$ equals the second derivative of the deflection $v''(\varphi)$. This approximation is valid up to $\alpha = 15° \mathrel{\hat=} 0.131\,\mathrm{rad}$, as $\arctan(0.131) = 0.130$. Substituting $1/R = -v''(\varphi)$ into the bending moment definition according to Equation (A.5), and solving for $v''(\varphi)$ gives

$$v''(\varphi) = -\frac{M(\varphi)}{E_0\ I}. \tag{A.26}$$

The integration of Equation (A.26), with the fixed support boundary conditions $v'(\varphi = \alpha/2) = 0$ and $v(\varphi = \alpha/2) = 0$, gives the radial deflection according to Equation (A.27).

$$v(\varphi) = -\frac{q_\mathrm{r}R^4}{2E_0 I}\left[-\frac{\cos\varphi}{\sin(\alpha/2)} - \frac{\varphi^2}{\alpha} + \frac{\cos(\alpha/2)}{\sin(\alpha/2)} + \frac{\alpha}{4}\right] \cdot \left[\frac{1}{g_1(\alpha) + g_2(\alpha)\cdot AR^2/I}\right] \tag{A.27}$$

The maximum deflection v_max occurs at $\varphi = 0$ in the beam midsection:

$$v_\mathrm{max} = -\frac{q_\mathrm{r}R^4}{2E_0 I}\left[-\frac{1}{\sin(\alpha/2)} + \frac{\cos(\alpha/2)}{\sin(\alpha/2)} + \frac{\alpha}{4}\right] \cdot \left[\frac{1}{g_1(\alpha) + g_2(\alpha)\cdot AR^2/I}\right] \tag{A.28}$$

Both normal stress $\sigma_\mathrm{n}(\varphi)$ and bending stress $\sigma_\mathrm{M}(\varphi)$ are defined by the normal force and bending moment according to

$$\sigma_\mathrm{n}(\varphi) = F_\mathrm{n}(\varphi)/A \tag{A.29}$$

$$\sigma_\mathrm{M}(\varphi) = M(\varphi)c/I, \tag{A.30}$$

where c represents the distance from the neutral surface, as visualized in Figure A.1. According to [11, p. 229], the neutral surface passes through the centroid of each structure, if it is subjected to pure bending, and if the stress is purely elastic. The bending stress is zero on the neutral surface, and reaches its maximum on the outer beam surfaces lying at $c = \pm t_\mathrm{s}/2$. In contrast, the normal stress is constant for a given cross-sectional surface at a specific angle φ.

This implies that also the strain splits into normal strain $\epsilon_\mathrm{n}(\varphi)$ and bending strain $\epsilon_\mathrm{M}(\varphi)$. The underlying equations result from Hook's law $\epsilon = \sigma/E_0$:

$$\epsilon_\mathrm{n}(\varphi) = \sigma_\mathrm{n}(\varphi)/E_0 \tag{A.31}$$

$$\epsilon_\mathrm{M}(\varphi) = \sigma_\mathrm{M}(\varphi)/E_0 \tag{A.32}$$

Figure A.4 visualizes the derived values $F_{\mathrm{n}}(\varphi)$, $M(\varphi)$, $v(\varphi)$, $\sigma_{\mathrm{n}}(\varphi)$, $\sigma_{\mathrm{M}}(\varphi)$, $\epsilon_{\mathrm{n}}(\varphi)$ and $\epsilon_{\mathrm{M}}(\varphi)$ for a conductor with the following parameters: $R = 700\,\mathrm{mm}$, $\alpha = 10°$, $h_{\mathrm{s}} = 10\,\mathrm{mm}$, $t_{\mathrm{s}} = 1.5\,\mathrm{mm}$, $E_0 = 110\,\mathrm{GPa}$ and $q_{\mathrm{r}} = 600\,\mathrm{N\,m^{-1}}$.
Due to symmetry issues the plots focus on the angular range of $0 < \varphi \leq \alpha/2$. The normal force is almost constant over the conductor length and is of compressive nature, as Figure A.4 a) states. The radial displacement v shown in Figure A.4 c) reveals a maximum displacement in the beam center of 0.258 mm. Figure A.4 d) plots the functions of stress and strain. The red curves correspond to the outer beam surface ($c = +t_{\mathrm{s}}/2$), while the blue curves represent the inner beam surface ($c = -t_{\mathrm{s}}/2$). Solid curves correspond to the total values which result from a summation of normal and bending strain or stress. Negative strain and stress are of compressive nature, while the positive sign corresponds to the tensile direction. For the given example the largest values occur at the beam end on the inner surface, with a compressive strain of $-6.1 \cdot 10^{-4}\mathrm{m/m}$, and a compressive stress of $-67.15\,\mathrm{MPa}$.

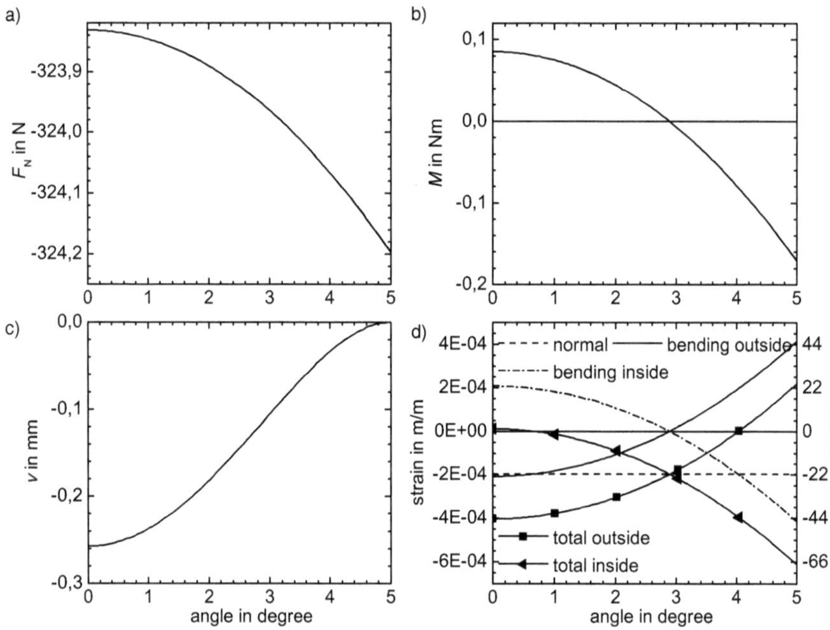

Figure A.4: Results for the sample beam subjected to a radial line load
a) Normal force, b) bending moment, c) radial displacement
d) strains and stresses on outer and inner beam surface

A.5 Correlation Between Tensile Strength and DP

The author's publication [84] contains an analytic correlation between DP and zero-span tensile index TI for accelerated aged insulating paper. The tensile index is proportional to the ultimate tensile strength σ_u according to

$$\sigma_\mathrm{u} = TI \cdot g/d_\mathrm{p}, \tag{A.33}$$

with paper grammage g and strip thickness d_p. Written in normalized form the correlation between normalized zero-span strength and DP deduced from [84] is given by

$$\bar{\sigma}_\mathrm{u} = 1.036 - 139.19/TI + 9.946 \cdot 10^{-3} \ln(TI). \tag{A.34}$$

Figure A.5 visualizes this correspondence. For factory new paper with $DP = 1300$ the normalized strength is 100 %. The case of 50 % of strength reduction corresponds to $DP = 223$, and a reduction to 20 % is characterized by $DP = 155$. Other authors came to similar relationships [25, 20]. The authors of [20] concluded that $DP = 200$ correlates to a decrease in zero-span strength to 20 percent of the initial strength.

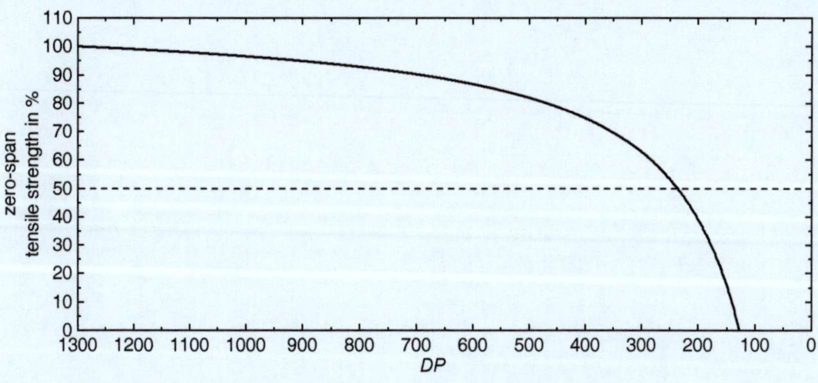

Figure A.5: Correlation between DP and normalized zero-span tensile strength

Appendix B

Tables and Datasheets

Table B.1: Ramberg-Osgood parameters for generic copper materials between $0 < \epsilon \leq \epsilon_{0.2}$

label	E_0 in GPa	$\sigma_{0.2}$ in MPa	k_R	m
co-65	130	65	3/7	4.6
co-90	110	90	3/7	5
co-180	110	180	3/7	5

Table B.2: Specification of tested CTCs

	h_s in mm	t_s in mm	d_p in µm	b_p in mm	N_p	p in mm	epoxy coated	k_f
CTC-A-11	5.70	1.10	80	30	13	1.80	yes	0.30
CTC-B-21	8.30	1.40	80	30	13	1.80	no	0.08
CTC-C-29	3.42	1.99	80	30	12	1.80	no	0.08
CTC-D-37	5.70	1.13	80	30	13	1.80	yes	0.30
CTC-E-19	5.15	1.50	100	25	4	0.80	no	0.08

KISTLER
measure. analyze. innovate.

ACCELERATION CALIBRATION CERTIFICATE

Type		**8772A50**	**Manufacturer**		**Kistler**
Serial Number		**2135730**	**Certificate ID #**		**41-2135730-130307T1055**
Time Constant	sec.	1.0	Mounted Resonant Frequency	kHz	20.0
Transverse Sensitivity	%	5.0	Bias Voltage	V_{DC}	11.7
Reference Specifications					
Range	*g*	± 50	Temp. Range, operating	°C	0 to 65
Max Range	*g*	± 80	Output Impedance	Ω	≤500

Frequency Response (Typical)

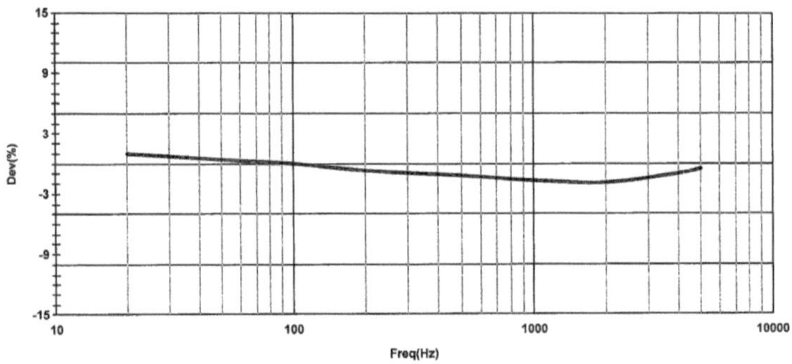

Sensitivity at 100 Hz, 3.0 *g* rms	mV/*g*	101.9		
Sensitivity at 159 Hz, 3.0 *g* rms	mV/*g*	101.4	*g* = 9.807 m/s²	159.2 Hz = 1000 rads/sec
Environmental Temperature	°C	22 ± 4	Condition	New
Relative Humidity	%	30 ± 30	NIST Test Report Number	683/282594-12
Calibration Date		3/7/2013	Calibration Technician:	Mark Thomas

This sensor was calibrated per Kistler test procedure 978-5355-701 using a comparison technique against a Kistler working standard. Kistler working standards are periodically calibrated against a primary standard system, which in turn is periodically recertified to the National Institute of Standards and Technology (NIST) or another recognized national standard. Measurements are derived from accepted values of natural physical constants according to the International System of Units (SI). This calibration meets or exceeds the requirements of MIL-STD-45662A, ISO 9001, ANSI/NCSL Z540-1 and is accredited to ISO/IEC 17025 as verified by the ANSI-ASQ National Accreditation Board/ACLASS. Refer to certificate and scope of accreditation AC-1117. Estimated uncertainty is ± 1.24% of reading with respect to the primary standard. Certificates are on file at Kistler and may be requested in writing. This certificate shall not be reproduced, except in full, without written approval of Kistler Instrument Corporation.

Reference Equipment	Manufacturer	Type	Serial Number	Reference Equipment	Manufacturer	Type	Serial Number
Accelerometer (Working Std.)	Kistler	8076K	C148323	Multimeter (Standard)	HP	34401A	3146A65537
Charge Amplifier (Working Std.)	Kistler	5020	C93919	Multimeter (Test)	HP	34401A	3146A65544
Accelerometer (Primary Std.)	Kistler	8002K	C197006	Function Generator	Wavetek	270	C6370761
Charge Amplifier (Primary Std.)	Kistler	5020	C4870	Charge Amplifier (Test)	Kistler	ETL1026	3

Kistler Instrument Corp.
75 John Glenn Drive
Amherst, NY 14228-2171

Tel 1-888-KISTLER
Fax 1-716-691-5226
info.us@kistler.com

ISO 9001 Certified Quality System
ISO 17025 Accredited Calibration Laboratory
040-0040-001 Rev B - Page 2 of 2

www.kistler.com

Figure B.1: Calibration certificate for the 50 g acceleration sensor

KISTLER

measure. analyze. innovate.

KALIBRIERSCHEIN	CALIBRATION CERTIFICATE
BESCHLEUNIGUNGSSENSOR	ACCELEROMETER

Typ / *Type*	8728A500	Hersteller / *Manufacturer*	Kistler
Serien-Nr / *Serial Number*	2120347	Cert ID # / *Cert ID #*	2120347-120629T0957

Bereich	*Range*	g	± 500
Max. Bereich	*Max. Range*	g	± 1000
Betriebstemperaturbereich	*Temp. Range, operating*	°C	-54 bis 120
Ausgangimpedanz	*Output Impedance*	Ω	≤100
Zeitkonstante	*Time Constant*	s	0,4
Seitenempfindlichkeit	*Transverse Sensitivity*	%	1,5
Ruhespannung	*Bias Voltage*	V_{DC}	10,4
...sonanzfrequenz montiert (fn)	*Mounted Resonant Frequency (fn)*	kHz	76,0

Typischer Frequenzgang — ***Typical Frequency Response***

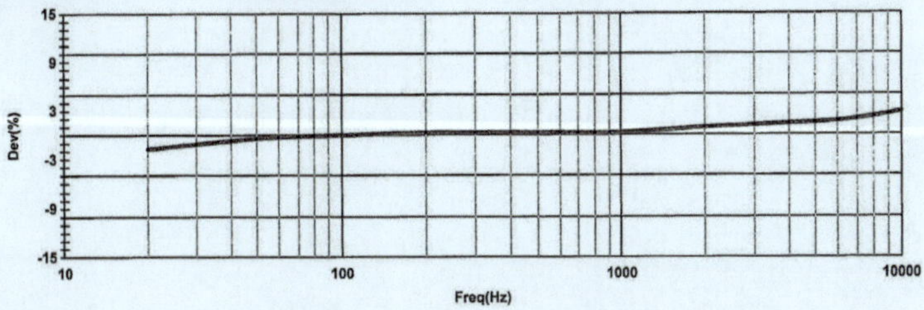

...mpfindlichkeit bei 159Hz, 10,0g rms	*Sensitivity at 159Hz, 10.0 g rms*	mV/g	10,43	*g = gn = 9,80665 m/s²*
Umgebungstemperatur	*Environmental Temperature*	°C	22 ± 4	
Relative Feuchte	*Relative Humidity*	%	30 ± 30	
Kalibriert von	*Calibration Technician*		Patrick Proctor	
Datum	*Calibration Date*		29.06.2012	

Dieser Sensor ist gemäss Kistler Testprozedur 978-5276-701 kalibriert worden, unter Verwendung einer Vergleichstechnik nach Kistler Arbeitsnormen. Kistler Arbeitsnormen werden periodisch mit einem primären Standardsystem abgeglichen, welches seinerseits periodisch mit dem National Institute of Standads and Technology (NIST) oder anderen anerkannten nationalen Standards in Einklang gebracht wird. Messungenwerden von anerkannten Werten physikalischer Konstanten abgeleitet, gemäss dem International System of Units (SI). Diese Kalibrierung erfüllt oder übersteigt die Anforderungen von MIL-STD-45662A, ISO 9001, ANSI/NCSL Z540-1 und entspricht ISO/IEC 17025, verifiziert durch das ANSI-ASQ National Accreditation Board/ACLASS . Siehe Zertifizierung AC-1117. Die geschätzte Ungenauigkeit beträgt ± 1.3% der gemessenen Werte gegenüber dem primären Standard. Zertifizierungen sind bei Kistler dokumentiert und können schriftlich angefordert werden. Dieses Zertifikat kann nicht ohne schriftliches Einverständnis von Kistler Instrumente AG, und nur in seiner Gesamtheit, reproduziert werden. *This sensor was calibrated per Kistler test procedure 978-5276-701 using a comparison technique against a Kistler working standard. Kistler working standards are periodically calibrated against a primary standard system, which in turn is periodically recertified to the National Institute of Standards and Technology (NIST) or another recognized national standard. Measurements are derived from accepted values of natural physical constants according to the International System of Units (SI). This calibration meets or exceeds the requirements of MIL-STD-45662A, ISO 9001, ANSI/NCSL Z540-1 and is accredited to ISO/IEC 17025 as verified by the ANSI-ASQ National Accreditation Board/ACLASS. Refer to certificate and scope of accreditation AC-1117. Estimated uncertainty is ± 1.3% of reading with respect to the primary standard. Certificates are on file at Kistler and may be requested in writing. This certificate shall not be reproduced, except in full, without written approval of Kistler Instrument Corporation.*

Referenz Geräte/*Reference*	Hersteller/*Mfg.*	Typ/*Type*	Serien-Nr/*SN*	Referenz Geräte/*Reference*	Hersteller/	Typ/*Type*	Serien-Nr/*SN*
Beschleunigungs-Sensor/*Accel.*(GN)	Kistler	809K112	2034904	Multimeter/*Multimeter* (UUT)	HP	34401A	US36022114
Ladungsverstärker/*Charge Amp.*(GN)	Kistler	5020	C31904	Multimeter/*Multimeter* (Std)	HP	34401A	US36021830
Beschleunigungs-Sensor/*Accel.*(BN)	Kistler	8002K	C139113	Funktionsgenerator/*Function Gen.*	Wavetek	270	D91050465
Ladungsverstärker/*Charge Amp.* (BN)	Kistler	5020	C92253	Ladungsverstärker/*Charge Amp.*(UUT)	Kistler	ETL1026	6

GN = Gebrauchsnormal/*Working Standard*, BN = Bezugsnormal/*Printory Standard*, UUT = Funktionsgenerator/*Unit Under Test*, Std = Normal/*Standard*

Kistler Instrument Corp.
75 John Glenn Drive
Amherst, NY 14228-2171

Tel 1-888-KISTLER
Fax 1-716-691-5226
info.us@kistler.com

ISO 9001 Certified Quality System
ISO 17025 Accredited Calibration Laboratory
040-0040-001 Rev B - Seite 2 von 2

www.kistler.com

Figure B.2: Calibration certificate for the 500 g acceleration sensor

Appendix C

Bibliography

C.1 References

[1] Hyun-Mo Ahn et al. “Finite-Element Analysis of Short-Circuit Electromagnetic Force in Power Transformer”. In: *IEEE Transactions on Industry Applications* 47.3 (May 2011), pp. 1267–1272. ISSN: 0093-9994. DOI: 10.1109/TIA.2011.2126031.

[2] Guillaume Aoust et al. “Viscous damping on flexural mechanical resonators”. In: *Sensors and Actuators A: Physical* 230 (July 2015), pp. 126–135. ISSN: 0924-4247. DOI: 10.1016/j.sna.2015.04.004.

[3] International Copper Association. “The copper advantage. A guide to working with copper and copper alloys”. In: *New York: International Copper Association LTD* (2011).

[4] Muneer Baig et al. “Shear and multiaxial responses of oxygen free high conductivity (OFHC) copper over wide range of strain-rates and temperatures and constitutive modeling”. In: *International Journal of Plasticity* 40 (2013), pp. 65–80. ISSN: 0749-6419. DOI: 10.1016/j.ijplas.2012.07.004.

[5] A. Bakshi. “An Investigation of Winding Curvature Effect on the Mechanical Strength of Transformer Windings”. In: *IEEE Transactions on Power Delivery* 30.4 (Aug. 2015), pp. 1821–1826. ISSN: 0885-8977. DOI: 10.1109/TPWRD.2015.2410775.

[6] A. Bakshi and S.V. Kulkarni. “Analysis of Buckling Strength of Inner Windings in Transformers Under Radial Short-Circuit Forces”. In: *IEEE Transactions on Power Delivery* 29.1 (Feb. 2014), pp. 241–245. ISSN: 0885-8977. DOI: 10.1109/TPWRD.2013.2272102.

[7] A. Bakshi and S.V. Kulkarni. "Coupled Electromagnetic-Structural Analysis of the Spiraling Phenomenon in a Helical Winding of a Power Transformer". In: *IEEE Transactions on Power Delivery* 29.1 (Feb. 2014), pp. 235–240. ISSN: 0885-8977. DOI: 10.1109/TPWRD.2013.2276047.

[8] Herbert Balke. *Einführung in die Technische Mechanik - Festigkeitslehre.* 3. Aufl. Berlin Heidelberg New York: Springer-Verlag, 2014. ISBN: 978-3-642-40981-3.

[9] Zdenek P. Bazant and Luigi Cedolin. *Stability of Structures: Elastic, Inelastic, Fracture and Damage Theories.* World Scientific Publishing Company, 2010. ISBN: 9814317039.

[10] R. Bedri and M. O. Al-Nais. "Numerical Analysis and Its Applications: Third International Conference, NAA 2004, Rousse, Bulgaria, June 29-July 3, 2004, Revised Selected Papers". In: ed. by Zhilin Li, Lubin Vulkov, and Jerzy Wasniewski. Berlin, Heidelberg: Springer Berlin Heidelberg, 2005. Chap. Prestressed Modal Analysis Using Finite Element Package ANSYS, pp. 171–178. ISBN: 978-3-540-31852-1. DOI: 10.1007/978-3-540-31852-1_19.

[11] Ferdinand P. Beer et al. *Mechanics of Materials.* 6. Ed. McGraw-Hill Education, 2011. ISBN: 0073380288.

[12] Giorgio Bertagnolli. *Short circuit duty of power transformers - The ABB approach.* 1996.

[13] T. Berther, J. Kubler, and G.H Gautschi. "K-Shear Accelerometers". In: *Noise and Vibration Worldwide.* Oxford Elsevier Advanced Technology. June 1997.

[14] S. S. Bhavikatti. *Finite Element Analysis.* New Delhi: New Age International, 2005. ISBN: 978-8-122-41589-6.

[15] Russell Cox et al. "Characteristics of laterally vibrating resonant microcantilevers in viscous liquid media". In: *Journal of Applied Physics* 111.1, 014907 (2012). ISSN: 0021-8979. DOI: 10.1063/1.3674278.

[16] Ercan M. Dede, Jaewook Lee, and Tsuyoshi Nomura. *Multiphysics Simulation - Electromechanical System Applications and Optimization.* Berlin, Heidelberg: Springer, 2014. ISBN: 978-1-447-15640-6.

[17] Robert M Del Vecchio et al. *Transformer design principles: with applications to core-form power transformers.* Boca Raton, Florida: CRC press, 2010.

[18] R. G. Dong. *Effective Mass and Damping of Submerged Structures.* Livermore, California: Lawrence Livermore Laboratory, University of California, 1978.

[19] Clive L Dym. *Stability theory and its applications to structural mechanics.* New York: Courier Corporation, 2012. ISBN: 978-0-486-14997-4.

[20] J Fabre and A Pichon. "Deterioration processes and products of paper ageing in oil applied to transformers". In: *Proceedings of CIGRE conference.* 1960, pp. 224–234.

[21] J. Faiz, B.M. Ebrahimi, and T. Noori. “Three- and Two-Dimensional Finite-Element Computation of Inrush Current and Short-Circuit Electromagnetic Forces on Windings of a Three-Phase Core-Type Power Transformer”. In: *IEEE Transactions on Magnetics* 44.5 (May 2008), pp. 590–597. ISSN: 0018-9464. DOI: 10.1109/TMAG.2008.917819.

[22] S.K. Gupta et al. “Critical aspects for short circuit performance of large power transformers connected to UHV systems”. In: *Proceedings of the 16th International Symposium on High Voltage Engineering.* .Paper F-9. SAIEE, Innes House. Johannesburg, 2009. ISBN: 978-0-620-44584-9.

[23] Markus Haid and Dieter Maier. *Verbesserung der referenzlosen inertialen Objektverfolgung zur low-cost Indoor-Navigation durch Anwendung der Kalman-Filterung.* Stuttgart: Fraunhofer-IRB-Verlag, 2005. ISBN: 978-3-816-76704-6.

[24] Martin J. Heathcote. *The J & P Transformer Book - A Practical Technology of the Power Transformer.* 0013. Aufl. London: Newnes, 2007. ISBN: 978-0-750-68164-3.

[25] R.J Heywood. “Degradation of cellulosic insulation in power transformers. Part 4 : Effects of ageing on the tensile strength of paper”. In: *IEE Proceedings Science, Measurement and Technology, Vol. 147, No. 6* (2000).

[26] S. L. Ho et al. “Numerical simulation of transient force and eddy current loss in a 720 MVA power transformer”. In: *IEEE Transactions on Magnetics* 40.2 (Mar. 2004), pp. 687–690. ISSN: 0018-9464. DOI: 10.1109/TMAG.2004.824801.

[27] ANSYS Inc. *ANSYS 16.2 Documentation.* SAS IP, Inc., 2015.

[28] J. T. P. de Castro J. G. T. Ribeiro and J. L. F. Freire. “New Improvement in the Digital Double Integration Filtering Method to Measure Displacements using Accelerometers”. In: *International Modal Analysis Conference* (2001).

[29] Jennifer L. Jordan et al. “Strain rate-dependant mechanical properties of OFHC copper”. In: *Journal of Materials Science* 48 (20 June 2013), pp. 7134–7141. ISSN: 1573-4803. DOI: 10.1007/s10853-013-7529-9.

[30] J.W. Morris Jr. “Dislocation-controlled Plasticity of Crystalline Materials: Overview”. In: *Encyclopedia of Materials: Science and Technology (Second Edition).* Ed. by K.H. Jürgen Buschow Robert W. Cahn Merton C. Flemings Bernhard Ilschner Edward J. Kramer Subhash Mahajan Patrick Veyssiere. Second Edition. Oxford: Elsevier, 2001, pp. 2245–2255. ISBN: 978-0-08-043152-9.

[31] Igor A Karnovsky. *Theory of Arched Structures.* Springer New York, 2012. ISBN: 978-1-4614-0469-9. DOI: 10.1007/978-1-4614-0469-9.

[32] Bernd Klein. *Leichtbau-Konstruktion - Berechnungsgrundlagen und Gestaltung.* Berlin Heidelberg New York: Springer-Verlag, 2013. ISBN: 978-3-658-02272-3.

[33] M Koch et al. "Onsite, Online and Post Mortem Insulation Diagnostics at Power Transformers". In: *Cigre SC A2 and D1 Colloquium Brugges: Belgium.* Vol. 23. 2007.

[34] H. Kojima et al. "Buckling Strength Analysis of Large Power Transformer Windings Subjected to Electromagnetic Force Under Short Circuit". In: *IEEE Transactions on Power Apparatus and Systems* PAS-99.3 (May 1980), pp. 1288–1297. ISSN: 0018-9510. DOI: 10.1109/TPAS.1980.319761.

[35] Mark Aleksandrovich Krasnoselskii et al. *Approximate solution of operator equations.* Springer Science and Business Media, 2012.

[36] S. V. Kulkarni and S. A. Khaparde. *Transformer Engineering - Design and Practice.* Boca Raton, Florida: CRC Press, 2004. ISBN: 978-0-824-75728-1.

[37] G. B. Kumbhar and S. V. Kulkarni. "Analysis of Short-Circuit Performance of Split-Winding Transformer Using Coupled Field-Circuit Approach". In: *IEEE Transactions on Power Delivery* 22.2 (Apr. 2007), pp. 936–943. ISSN: 0885-8977. DOI: 10.1109/TPWRD.2007.893442.

[38] K. Kurita et al. "Mechanical Strength of Transformer Windings under Short-Circuit Conditions". In: *IEEE Transactions on Power Apparatus and Systems* PAS-88.3 (Mar. 1969), pp. 222–230. ISSN: 0018-9510. DOI: 10.1109/TPAS.1969.292310.

[39] Horace Lamb. "On the Flexure and the Vibrations of a Curved Bar". In: *Proceedings of the London Mathematical Society* 19.1 (1887), pp. 365–377. DOI: 10.1112/plms/s1-19.1.365.

[40] Q. Liu et al. "Assessment of ageing conditions through paper tensile strength analysis of scrapped transformers". In: *International Symposium on High Voltage Engineering (ISH)* (Aug. 2009), pp. 264–269.

[41] Erdogan Madenci and Ibrahim Guven. *The Finite Element Method and Applications in Engineering Using ANSYS.* Springer Science and Business Media, 2015. DOI: 10.1007/978-1-4899-7550-8.

[42] Leszek Majkut. "Free and forced vibrations of Timoshenko beams described by single difference equation". In: *Journal of Theoretical and Applied Mechanics* 47.1 (2009), pp. 193–210.

[43] W.J. McNutt et al. "Transformer short-circuit strength and standards 2014;A state-of-the-art paper". In: *Power Apparatus and Systems, IEEE Transactions on* 94.2 (Mar. 1975), pp. 432–443. ISSN: 0018-9510. DOI: 10.1109/TPAS.1975.31870.

[44] V Mollon et al. "Numerical Analsysis of Delamination growth in composite materials using two step extension and cohesive zone methods". In: *ECCM15 - 15th European Conference on Composite Materials* (2012).

[45] H. P. Moser and V. Dahinden. *Transformerboard.* Birkhäuser AG, 1979.

[46] A Needleman. “An analysis of intersonic crack growth under shear loading”. In: *Journal of Applied Mechanics* 66.4 (1999), pp. 847–857. DOI: 10.1115/1.2791788.

[47] R. Obika et al. “Development of magnet wire having excellent windability”. In: *Electrical Insulation Conference and Electrical Manufacturing & Coil Winding Conference, 1999. Proceedings.* 1999, pp. 479–483. DOI: 10.1109/EEIC.1999.826257.

[48] Y. H. Oh et al. “Displacement measurement of high-voltage winding for design verification of short-circuit strength of transformer”. In: *Proc. IEEE PES Transmission and Distribution Conf. and Exposition.* Vol. 2. 2003, pp. 831–835.

[49] M.R. Patel. “Instability of the continuously transposed cable under axial short-circuit forces in transformers”. In: *IEEE Transactions on Power Delivery* 17.1 (Jan. 2002), pp. 149–154. ISSN: 0885-8977. DOI: 10.1109/61.974202.

[50] Carlos G. Davila Pedro Camanho. “Mixed-Mode Decohesion Finite Elements for the Simulation of Delamination in Composite Materials”. In: *J. of Composite Materials* 37 (2003), pp. 1415–1438.

[51] Arne Petersen. *The use of Paperless CTC in Power Transformers.* Tech. rep. Cigre Study Commitee A2 Transformers, June 2003.

[52] Andre Preumont. *Mechatronics: Dynamics of Electromechanical and Piezoelectric Systems.* Vol. 136. Springer Science and Business Media, 2006.

[53] Thomas A Prevost. “Maintaining Short Circuit Strength in Transformers”. In: *Weidmann-ACTI Second Annual Technical Conference.* 2003.

[54] J. Wildeboer R. Boersma. “The Short-Circuit Strength of the inner windings of transformers against radial forces”. In: *Conference Internationale des Grands Reseaux Electriques a Haute Tension* (May 1962).

[55] Singiresu S. Rao. *The Finite Element Method in Engineering.* Oxford: Butterworth Heinemann, 2011. ISBN: 978-0-080-47050-4.

[56] Kim J.R. Rasmussen. “Full range stress strain curves for stainless steel alloys”. In: *Journal of Constructional Steel Research* 59 (2003), pp. 47–61.

[57] M.P. Saravolac et al. “Design Verification Criteria for evaluating the Short Circuit Withstand Capability of Transformer Inner Windings”. In: *CIGRE 2000* 12-208 (2000).

[58] Howard L Schreyer and Ernest F Masur. “Buckling of shallow arches”. In: *Journal of the Engineering Mechanics Division* 92.4 (1966), pp. 1–20.

[59] S. Tenbohlen. “Trends of the Diagnosis of HighVoltage Equipment”. In: ETG Kongress Fulda. Nov. 2012.

[60] F. Tillery Thompson H.A. and D. U. von Rosenberg. "The Dynamic Response of Low Voltage, High Current Disk Type Transformer Windings to Through Fault Loads". In: *IEEE Transactions on Power Apparatus and Systems* (1979), pp. 1091–1098.

[61] C. H. Thornton, S. Harper, and J. E. Bowers. "A Critical Survey of Available High-Temperature Mechanical Property Data for Copper and Copper Alloys". In: *International Copper Research Assoc., Inc., 1983,* (1983). Ed. by New York International Copper Research Association Inc. 708 Third Avenue.

[62] Johan Tryding. "In-Plane Fracture of Paper". eng. PhD thesis. Lund Institute of Technology, Division of Structural Mechanics, 1996, p. 148.

[63] A. Turon et al. "An engineering solution for mesh size effects in the simulation of delamination using cohesive zone models". In: *Engineering Fracture Mechanics* 74.10 (2007), pp. 1665–1682. ISSN: 0013-7944. DOI: 10.1016/j.engfracmech.2006.08.025.

[64] Wei-Chau Xie. *Dynamic Stability of Structures.* Cambridge: Cambridge University Press, 2006. ISBN: 978-0-521-85266-1.

[65] Haijun Zhang et al. "Dynamic Deformation Analysis of Power Transformer Windings in Short-Circuit Fault by FEM". In: *IEEE Transactions on Applied Superconductivity* 24.3 (June 2014), pp. 1–4. ISSN: 1051-8223. DOI: 10.1109/TASC.2013.2285335.

[66] X. Zhang and E. Gockenbach. "Asset-Management of Transformers Based on Condition Monitoring and Standard Diagnosis". In: *IEEE Electrical Insulation Magazine* 24.4 (July 2008), pp. 26–40. ISSN: 0883-7554. DOI: 10.1109/MEI.2008.4581371.

C.2 Standards

[67] DIN 50125:2009. *Prüfung metallischer Werkstoffe - Zugproben.* July 2009.

[68] IEC 60076-5 (2006-02) Ed. 3.0. *Power transformers - Part 5: Ability to withstand short circuit.* Feb. 2006.

[69] ASTM D1002-10:2010. *Standard Test Method for Apparent Shear Strength of Single-Lap-Joint Adhesively Bonded Metal Specimens by Tension Loading (Metal-to-Metal).* 2010.

[70] ASTM D5804-97:2002. *Standard Test Methods for Zero-Span Tensile Strength (Dry Zero-Span Tensile).* 2002.

[71] DIN EN 60317-0-2:2014-10. *Technische Lieferbedingungen für bestimmte Typen von Wickeldrähten - Teil 0-2: Allgemeine Anforderungen - Lackisolierte Flachdrähte aus Kupfer.* Oct. 2014.

[72] DIN EN ISO 11339:2010. *Klebstoffe - T-Schälprüfung für geklebte Verbindungen aus flexiblen Fügeteilen.* June 2010.

[73] DIN EN ISO 1924-2:2009. *Papier und Pappe - Bestimmung von Eigenschaften bei Zugbeanspruchung - Teil 2: Verfahren mit konstanter Dehngeschwindigkeit (20 mm/min)*. May 2009.

[74] DIN EN ISO 6892-1:2009. *Metallische Werkstoffe - Zugversuch - Teil 1: Prüfverfahren bei Raumtemperatur*. Dec. 2009.

[75] DIN EN ISO 7438:2012. *Metallische Werkstoffe - Biegeversuch*. Mar. 2012.

[76] EN ISO 15361:2000. *Pulps - Determination of zero-span tensile strength, wet or dry*. 2000.

[77] DIN EN ISO 178:2013. *Kunststoffe - Bestimmung der Biegeeigenschaften*. Sept. 2013.

[78] DIN ISO 5348:1999-07. *Mechanische Schwingungen und Stöße - Mechanische Ankopplung von Beschleunigungsaufnehmern*. July 1997.

[79] TAPPI T 231:1996. *Zero-span breaking strength of pulp (dry zero-span tensile)*. 1996.

C.3 Author's Journal Publications

[80] D. Geissler, C. Geissler, and T. Leibfried. "Simplified simulation model of continuously transposed cable for linear and nonlinear buckling analysis". In: *Mathematics and Computers in Simulation* (June 26, 2015). ISSN: 0378-4754. DOI: 10.1016/j.matcom.2015.06.005.

[81] D. Geissler and T. Leibfried. "Short Circuit Strength of Power Transformer Windings - Verification of Tests by a Finite Element Analysis Based Model". In: *IEEE Transactions on Power Delivery* (May 21, 2016). DOI: 10.1109/TPWRD.2016.2572399.

[82] M. Jaya, D. Geissler, and T. Leibfried. "Accelerating Dielectric Response Measurements on Power Transformers - Part I: A Frequency-Domain Approach". In: *IEEE Transactions on Power Delivery* 28.3 (Apr. 4, 2013), pp. 1469–1473. ISSN: 0885-8977. DOI: 10.1109/TPWRD.2013.2253135.

C.4 Author's Conference Proceedings

[83] D. Geissler, C. Geissler, and T. Leibfried. "Linear Buckling Analysis Model for Continuously Transposed Conductors Using Mechanical FEA". In: *11th International Conference on Modeling and Simulation of Electric Machines, Converters and Systems (ELECTRIMACS 2014)*. Universitat Politecnica de Valencia, May 19, 2014.

[84] D. Geissler, S. Gorges, and T. Leibfried. "Analysis of Youngs Modulus, Zero- and Wide-Span Tensile Index and Degree of Polymerization of Insulating Paper for Power Transformers from Accelerated Aging Tests". In: *Diagnostik elektrischer Betriebsmittel 2014 (ETG-FB 144)*. Nov. 25, 2014. ISBN: 978-3-8007-3648-5.

[85] D. Geissler, M. Jaya, and T. Leibfried. "Analysis of frequency domain spectroscopy measurements on power transformers by the use of a Finite Element based model". In: *Condition Monitoring and Diagnosis (CMD), 2012 International Conference on*. Sept. 23, 2012, pp. 206–210. DOI: 10.1109/CMD.2012.6416411.

[86] D. Geissler and T. Leibfried. "Test Setup and Results concerning the mechanical strength of Continuously Transposed Conductors for Power Transformers". In: *18th International Symposium on High Voltage Engineering*. Hanyang University Seoul Korea, Aug. 25, 2012, pp. 592–597. ISBN: 978-89-861-0195-9.

[87] D. Geissler and T. Leibfried. "Versuchsaufbau und Simulationen zur Bestimmung der mechanischen Festigkeit von Drillleitern fuer Leistungstransformatoren". In: *Diagnostik elektrischer Betriebsmittel 2012 (ETG-FB 134)*. Nov. 15, 2012.

[88] D. Geissler and T. Leibfried. "Mechanical Simulations Regarding the Influence of Paper Insulation Degradation on the Radial Mechanical Strength of Continuously Transposed Conductors for Power Transformers". In: *Nordic Insulation Symposium*. Akademika Forlag, June 21, 2013, pp. 103–107. ISBN: 9788232102747. URL: http://www.detgarbra.no/publish/fil/vis/1257.

[89] D. Geissler and T. Leibfried. "Deformation measurement system for buckling tests on Continuously Transposed Cable using a high-speed camera". In: *Optimization of Electrical and Electronic Equipment (OPTIM), 2014 International Conference on*. May 22, 2014, pp. 111–117. DOI: 10.1109/OPTIM.2014.6850895.

[90] D. Geissler and T. Leibfried. "Finite Element Analysis Model for Failure Analysis of Epoxy-Bonded Continuously Transposed Cable". In: *19th International Symposium on High Voltage Engineering (ISH 2015)*. 141. Aug. 23, 2015. ISBN: 978-80-261-0477-3.

[91] D. Geissler and T. Leibfried. "Mechanical breakdown of aged insulating paper around continuously transposed conductors for power transformers under the influence of short-circuit forces - analysis by numerical simulations". In: *Electrical Insulation Conference (EIC), 2015 IEEE*. June 7, 2015, pp. 401–406. DOI: 10.1109/ICACACT.2014.7223476.

[92] D. Geissler, T. Sosnik, and T. Leibfried. "Lineare und nichtlineare Buckling-Analysen an Drillleiter-Wicklungen fuer Leistungstransformatoren". In: *ANSYS Conference und 32. CADFEM Users' Meeting 2014*. June 4, 2014. ISBN: 3-937523-11-1.

C.5 Supervised Theses Contributing to This Work

[93] Simon Gorges. “Alterungsabhängige Charakterisierung von Isolierpapier für Leistungstransformatoren mittels Wide- und Zero-Span Zugversuchen”. Bachelor’s Thesis. Karlsruhe Institute of Technology, 2014.

[94] Nico Hübner. “Messung und Bewertung des Einflusses der Papierisolation auf die Biegesteifigkeit von Transformator-Drillleitern”. Bachelor’s Thesis. Karlsruhe Institute of Technology, 2014.

[95] Benjamin Klaus. “Entwicklung der kombinierten Ansteuerung von Leistungs- und Thyristorschalter für das phasenbezogene Schalten eines Hochstromtransformators für Kurzschlussversuche”. Diploma Thesis. Karlsruhe Institute of Technology, 2012.

[96] Jan Letsch. “Qualifizierung und Aufbau eines Messsystems zur Deformationsmessung an Transformator Drillleitern unter dem Einfluss von Stromkräften”. Bachelor’s Thesis. Karlsruhe Institute of Technology, 2013.

[97] Konstantin Meng. “Gekoppelte magnetisch-mechanische Simulation des Knickverhaltens von Transformator-Leitern”. Diploma Thesis. Karlsruhe Institute of Technology, 2015.

[98] Thomas Sosnik. “Entwicklung von Simulationsmodellen zum Knickverhalten von Drillleitern für Leistungstransformatoren”. Diploma Thesis. Karlsruhe Institute of Technology, 2014.

Nomenclature

α	Circumferential angle of conductor sector	°
α_e	Angle of camera y-axis alignment	°
$\bar{\epsilon}_{0.2}$	Transformed strain for Ramberg-Osgood expression	1
$\bar{\sigma}_{0.2}$	Transformed stress for Ramberg-Osgood expression	Pa
$\bar{B}_{z,LV}$, $\bar{B}_{z,HV}$	Average axial leakage flux through LV (HV) winding	T
ψ	Eigenvector	1
Δf	Fourier Transformation frequency resolution	Hz
ΔL	Member elongation	m
δ	Substituting parameter for Mathieu equation	1
δ_s	Strain rate amplitude	$\frac{1}{s}$
ϵ	Strain	1
ϵ_0	Ramberg-Osgood parameter	1
ϵ_e	Engineering strain	1
ϵ_{lat}, ϵ_{long}	Lateral (longitudinal) strain	1
ϵ_{max}	Maximum strain	1
ϵ_M	Bending strain	1
ϵ_n	Normal strain	1
ϵ_t	True strain	1
ϵ_u	Ultimate strain at sample break	1

$\epsilon_{0.2}$	Strain at yield point	1
γ_{s}	Shearing angle	1
$\hat{I}_{\mathrm{k}}$	Peak of symmetrical short-circuit current	A
$\hat{I}_{\mathrm{max}}$	Maximum asymmetrical short-circuit current	A
$\hat{q}_{\mathrm{r}}$	Amplitude of time dependent radial line load	$\frac{\mathrm{N}}{\mathrm{m}}$
$\hat{v}_{\mathrm{w}}$	Peak radial winding marker movement	m
$\hat{V}_n$	Total max. radial conductor deformation after test no. n	m
$\hat{v}_n$	Max. radial conductor deformation at test no. n	m
κ	Coefficient	1
λ_j	Load multiplier for eigenvalue buckling analysis	1
$\mathbf{A}$	Magnetic vector potential vector	$\frac{\mathrm{Vs}}{\mathrm{m}}$
$\mathbf{B}$	Magnetic flux density vector	T
$\mathbf{C}$	Elasticity matrix	Pa
$\mathbf{c}$	Marker tracking coordinate vector	m
$\mathbf{E}$	Electric field vector	$\frac{\mathrm{V}}{\mathrm{m}}$
$\mathbf{F}$	Electromagnetic or exciting force vector	N
$\mathbf{f}$	Electromagnetic force density vector	$\frac{\mathrm{N}}{\mathrm{m}^3}$
$\mathbf{F}_{\mathrm{e,m}}$	Element magnetic force vector	1
$\mathbf{f}_{\mathrm{s}}$	Surface force vector	$\frac{\mathrm{N}}{\mathrm{m}}^2$
$\mathbf{f}_{\mathrm{v}}$	Volume force vector	$\frac{\mathrm{N}}{\mathrm{m}^3}$
$\mathbf{H}$	Magnetic field vector	$\frac{\mathrm{A}}{\mathrm{m}}$
$\mathbf{I}$	Current vector	A
$\mathbf{J}$	Current density vector	$\frac{\mathrm{A}}{\mathrm{m}^2}$
$\mathbf{K}$	Global system matrix	$\frac{\mathrm{N}}{\mathrm{m}}$
$\mathbf{K}_{\mathrm{e,m}}$	Element magnetic stiffness matrix	1
$\mathbf{N}$	Vector of interpolation functions	m

$\mathbf{P}$	Strain displacement matrix	$\frac{1}{\mathrm{m}}$
$\mathbf{q}$	Line load vector	$\frac{\mathrm{N}}{\mathrm{m}}$
$\mathbf{S}$	Stress stiffness matrix	Pa
$\mathbf{u}$	Displacement vector	m
$\mathbf{u}_\mathrm{n}$	Nodal displacement vector	m
$\mathbf{v}$	Velocity vector	$\frac{\mathrm{m}}{\mathrm{s}}$
μ	Permeability	$\frac{\mathrm{N}}{\mathrm{A}^2}$
μ_0	Vacuum permeability	$\frac{\mathrm{N}}{\mathrm{A}^2}$
μ_f	Dynamic viscosity	$\frac{\mathrm{kg}}{\mathrm{ms}}$
ν	Normalized parameter for Mathieu equation	1
ν_p	Poisson's ratio	1
ω	Angular frequency	Hz
ω_0	Power grid angular frequency	Hz
ω_cr	Critical angular frequency	Hz
Π	Potential energy	J
Π_e	Potential element energy	J
ψ	Normalized parameter for Mathieu equation	1
ρ_c	Space charge density	$\frac{\mathrm{As}}{\mathrm{m}^3}$
ρ_f, ρ_co	Fluid (copper) density	$\frac{\mathrm{kg}}{\mathrm{m}^3}$
σ	Stress	Pa
σ_0	Ramberg-Osgood parameter	Pa
σ_cr	Critical hoop stress	Pa
σ_e	Engineering stress	Pa
σ_max	Maximum stress	Pa
σ_M	Bending stress	Pa
$\sigma_\mathrm{n,max}$, $\sigma_\mathrm{t,max}$	Normal (tangential) contact stress	Pa

σ_{n}	Normal stress	Pa
σ_{t}	True stress	Pa
σ_{u}	Ultimate tensile strenght at sample break	Pa
$\sigma_{0.2}$	Stress at yield point	Pa
τ	Substituting parameter for Mathieu equation	1
τ_{k}	Circuit time constant	s
τ_{max}	Maximum shearing stress	Pa
τ_{s}	Shearing stress	Pa
$\tilde{\mathbf{c}}$	Marker tracking relative coordinates vector	m
$\tilde{\mathbf{c}}_{\mathrm{w,c}}$	Compensated, relative coordinates of winding marker	m
$\tilde{\mathbf{c}}_{\mathrm{w}}$, $\tilde{\mathbf{c}}_{\mathrm{ref}}$	Relative coordinates of winding (reference) marker	m
$\tilde{\mathbf{u}}$	Approximated displacement vector	m
$\tilde{\sigma}_0$	Proportionality limit (Ramberg-Osgood expression)	Pa
$\tilde{h}_{\mathrm{s}}$	Reduced strand height	m
$\tilde{m}$	Ramberg-Osgood parameter	1
$\tilde{T}$	Scaled slenderness ratio	1
$\tilde{y}_{\mathrm{w,c}}$	Compensated, relative y-coordinate of winding marker	m
Λ	Roots of characteristic equation	1
φ	Azimuthal cylindrical coordinate	$^\circ$
φ_{k}	Circuit impedance phase angle	$^\circ$
ϑ	Feeding voltage phase angle	$^\circ$
ξ	Half number of axial supporting sticks	1
A	Cross-sectional area	m^2
A_1, A_2, A_3	Coefficients	m
A_{o}	Overlapping area	m^2
a_i	Even Fourier parameter no. i	m

A_j	Conductor no. j cross-sectional area	m^2
B_{duct}	Magnetic flux density in the main duct	T
b_{p}	Paper strip width	m
B_{r}, B_{φ}, B_{z}	Radial (azimuthal, axial) magnetic flux density	T
b_i	Odd Fourier parameter no. i	m
c	Distance from neutral surface inside a beam	m
$d\varphi$	Angle of bending	°
D_{m}	Mean winding diameter	m
d_{p}	Paper strip thickness	m
DP	Degree of polymerization	1
ds, ds^*	Arc length inside beam before (after) bending	m
e	Relative error	1
E_0	Young's modulus	Pa
$E_{0.2}$	Young's modulus for Ramberg-Osgood expression	Pa
E_{pap}	Homogenized Young's modulus of paper insulation	Pa
E_{t}	Tangent modulus of elasticity	Pa
F	Force	N
$f(t)$	Function for Bubnov-Galerkin approach	m
f_0	Power grid frequency	Hz
f_1	Lowest characteristic buckling frequency	Hz
F_{cr}	Buckling force	N
f_{cr}	Critical frequency	Hz
F_{n}	Normal force	N
F_{p}	Perturbation force	N
F_{s}	Reaction force at supports	N
F_{u}	Ultimate tensile force at sample break	N

$f_{\mathrm{e,m},ij}$	Entry of element magnetic force vector	N
G	Shear modulus	Pa
g	Grammage of insulating paper	$\frac{\mathrm{g}}{\mathrm{m}^2}$
g_1, g_2	Abbreviating functions	1
G_n, G_t	Normal (tangential) energy release rate	$\frac{\mathrm{J}}{\mathrm{m}^2}$
h_c	Height of CTC copper core	m
h_duct	Height of main duct	m
h_s	Strand height	m
h_r	Rise of curved winding sector	m
I	Area moment of inertia	m^4
i	Complex Unit	1
$i_\mathrm{k}(t)$	Short-circuit current	A
I_pap	Area moment of inertia of paper insulation	m^4
I_r, I_φ, I_z	Radial (azimuthal, axial) current	A
j	Index	1
K	Contact stiffness	Nm^3
k	Index	1
k_f	Frictional coefficient	1
k_m	Hydrodynamic mass coefficient	$\frac{\mathrm{kg}}{\mathrm{m}}$
K_n, K_t	Normal (tangential) contact stiffness	Nm^3
k_q	Coefficient	1
k_R	Ramberg-Osgood parameter	1
k_sl	Slenderness factor	1
$k_{\mathrm{e,m},ij}$	Entry of element magnetic stiffness matrix	$\frac{\mathrm{N}}{\mathrm{m}}$
$k_{\mathrm{r},j}$, $k_{\mathrm{z},j}$	Radial (axial) load factor for winding no. j	$\frac{\mathrm{N}}{\mathrm{mA}^2}$
L	Length of conductor (member, beam) support distance	m

L_{LV}, L_{HV}	Number of LV (HV) winding layers	1
M	Bending moment	Nm
m	Ramberg-Osgood parameter	1
M_0	Bending moment in conductor midsection	Nm
m_{c}	Conductor mass per unit length	$\frac{\mathrm{kg}}{\mathrm{m}}$
N	Number of points of Fourier Transformation	1
n	Index	1
N_{as}	Number of axial supporting sticks	1
n_{c}	Filter cut of index	1
N_{LV}	Number of LV winding turns	1
N_{p}	Number of paper layers	1
N_{s}	Number of strands inside CTC	1
N_i	Interpolation function for finite element node no. i	m
p	Nominal increase due to paper insulation	m
p_{p}	Perturbation pressure	Pa
Q	Electric charge	As
q_{cr}	Critical line load	$\frac{\mathrm{N}}{\mathrm{m}}$
$q_{\mathrm{r,LV}}$, $q_{\mathrm{r,HV}}$	Radial line load on conductor inside LV (HV) winding	$\frac{\mathrm{N}}{\mathrm{m}}$
q_{r}, q_{φ}, q_{z}	Radial (azimuthal, axial) line load	$\frac{\mathrm{N}}{\mathrm{m}}$
$q_{\mathrm{r},j}$	Radial line load acting on winding no. j	$\frac{\mathrm{N}}{\mathrm{m}}$
R	Radius	m
r	Radial cylindrical coordinate	m
$R_{\mathrm{p,0.2}}$	Yield point	Pa
r_{s}	Stiffness ratio	1
R_{t}, R_{g}, R_{k}	Transformer (grid, short-circuit) resistance	Ω
Re	Reynolds number	1

S	Bending stiffness	Nm^2
S_{p}, S_{co}, S_{pap}	CTC (copper core, paper covering) bending stiffness	Nm^2
T	Slenderness ratio	1
t	Time	s
t_{c}	Thickness of CTC copper core	m
t_{s}	Strand thickness	m
TI	Tensile index of insulating paper	$\frac{\mathrm{Nm}}{\mathrm{g}}$
U	Total strain energy	J
u	Tangential displacement	m
U_0	Feeding voltage amplitude	V
$u_0(t)$	Feeding voltage	V
v	Radial displacement	m
v_{g}	Radial g-sensor movement	m
v_{max}	Maximum radial displacement	m
v_{w}	Radial winding marker movement	m
$V_{\mathrm{r},n}$	Total residual conductor deformation after test no. n	m
$v_{\mathrm{r},n}$	Residual conductor deformation after test no. n	m
W	Complex spectrum	m
$w(t)$	Generic signal	m
$w_{\mathrm{f}}(t)$	Filtered signal	m
x	Maximum beam deflection	m
X_{t}, X_{g}, X_{k}	Transformer (grid, short-circuit) reactance	Ω
z	Axial cylindrical coordinate	m
CD	Cross Direction	
CTC	Continuously Transposed Conductor	
CZM	Cohesive Zone Model	

DP	Degree of Polymerization
EMC	Electromagnetic Compatibility
ETP	Electrolytic Tough Pitch (copper)
FEA	Finite Element Analysis
FFT	Fast-Fourier-Transformation
HP	High Pass (filter)
HV	High Voltage (winding)
HVDC	High Voltage Direct Current
IFFT	Inverse Fast-Fourier-Transformation
LV	Low Voltage (winding)
MD	Manufacturing Direction
OFHC	Oxygen-Free High Conductivity (copper)
RMS	Root Mean Square